INNOVATIVE SITE REMEDIATION TECHNOLOGY

CHEMICAL TREATMENT

One of an Eight-Volume Series

Edited by
William C. Anderson, P.E., DEE
Executive Director, American Academy of Environmental Engineers

1994

Prepared by WASTECH®, a multiorganization cooperative project managed by the American Academy of Environmental Engineers® with grant assistance from the U.S. Environmental Protection Agency, the U.S. Department of Defense, and the U.S. Department of Energy.

The following organizations participated in the preparation and review of this volume:

Air & Waste Management Association
P.O. Box 2861
Pittsburgh, PA 15230

American Society of Civil Engineers
345 East 47th Street
New York, NY 10017

American Academy of Environmental Engineers®
130 Holiday Court, Suite 100
Annapolis, MD 21401

Hazardous Waste Action Coalition
1015 15th Street, N.W., Suite 802
Washington, D.C. 20005

American Institute of Chemical Engineers
345 East 47th Street
New York, NY 10017

Water Environment Federation
601 Wythe Street
Alexandria, VA 22314

Library of Congress Cataloging in Publication Data

Innovative site remediation technology/ edited by William C. Anderson
200p. 15.24 x 22.86cm.
Includes bibliographic references.
Contents: -- [2] Chemical treatment [3] Soil washing/soil flushing
-- [4] Stabilization/solidification
-- [6] Themal desorption.
1. Soil remediation. I. Anderson, William, C., 1943-
II. American Academy of Environmental Engineers.
TD878.I55 1994 628.5'5 93-20786

Distributed by: Springer-Verlag Berlin Heidelberg New York
London Paris Tokyo Hong Barcelona Budapest
ISBN 3-540-58911-2 Springer-Verlag Berlin Heidelberg New York

Book design by Lori Imhoff

Printed in the United States of America

CONTRIBUTORS

This monograph was prepared under the supervision of the WASTECH® Steering Committee. The manuscript for the monograph was written by a task group of experts in chemical treatment and was, in turn, subjected to two peer reviews. One review was conducted under the auspices of the Steering Committee and the second by professional and technical organizations having substantial interest in the subject.

PRINCIPAL AUTHORS

Leo Weitzman, Ph.D., *Task Group Chair*
President
LVW Associates, Inc.

Kimberly Gray, Ph.D.
Assistant Professor
Department of Civil Engineering and Geological Sciences
University of Notre Dame

Robert W. Peters, Ph.D., P.E., DEE
Environmental Systems Engineer
Energy Systems Division
Argonne National Laboratory

Frederick K. Kawahara, Ph.D.
Research Chemist
Risk Reduction Engineering Laboratory
U.S. Environmental Protection Agency

John Verbicky, Ph.D.
Chemfab Corporation

REVIEWERS

The panel that reviewed the monograph under the auspices of the Project Steering Committee was composed of:

Peter B. Lederman, Ph.D., P.E., DEE, P.P., *Chair*
Center for Environmental Engineering and Science
New Jersey Institute of Technology

B. Mo Kim, Ph.D.
Environmental Laboratory
General Electric Company

John Herrmann
Senior Technical Advisor
Water and Hazardous Waste Treatment Research Division
U.S. Environmental Protection Agency

Joseph F. Lagnese, Jr., P.E., DEE
Private Consultant

STEERING COMMITTEE

Frederick G. Pohland, Ph.D., P.E., DEE
Chair
Weidlein Professor of Environmental Engineering
University of Pittsburgh

William C. Anderson, P.E., DEE
Project Manager
Executive Director
American Academy of Environmental Engineers

Paul L. Busch, Ph.D., P.E., DEE
President and CEO
Malcolm Pirnie, Inc.
Representing, American Academy of Environmental Engineers

Richard A. Conway, P.E., DEE
Senior Corporate Fellow
Union Carbide Corporation
Chair, Environmental Engineering Committee
EPA Science Advisory Board

Timothy B. Holbrook, P.E.
Engineering Manager
Groundwater Technology, Inc.
Representing, Air & Waste Management Association

Walter W. Kovalick, Jr., Ph.D.
Director, Technology Innovation Office
Office of Solid Waste and Emergency Response
U.S. Environmental Protection Agency

Joseph F. Lagnese, Jr., P.E., DEE
Private Consultant
Representing, Water Environment Federation

Peter B. Lederman, Ph.D., P.E., DEE, P.P.
Center for Env. Engineering & Science
New Jersey Institute of Technology
Representing, American Institute of Chemical Engineers

Raymond C. Loehr, Ph.D., P.E., DEE
H.M. Alharthy Centennial Chair and Professor
Civil Engineering Department
University of Texas

James A. Marsh
Office of Assistant Secretary of Defense for Environmental Technology

Timothy Oppelt
Director, Risk Reduction Engineering Laboratory
U.S. Environmental Protection Agency

George Pierce, Ph.D.
Editor in Chief
Journal of Microbiology
Manager, Bioremediation Technology Dev.
American Cyanamid Company
Representing the Society of Industrial Microbiology

H. Gerard Schwartz, Jr., Ph.D., P.E., DEE
Senior Vice President
Sverdrup Corporation
Representing, American Society of Civil Engineers

Claire H. Sink
Acting Director
Division of Technical Innovation
Office of Technical Integration
Environmental Education Development
U.S. Department of Energy

Peter W. Tunnicliffe, P.E., DEE
Senior Vice President
Camp Dresser & McKee, Incorporated
Representing, Hazardous Waste Action Coalition

Charles O. Velzy, P.E., DEE
Private Consultant
Representing, American Society of Mechanical Engineers

William A. Wallace
Vice President, Hazardous Waste Management
CH2M Hill
Representing, Hazardous Waste Action Coalition

Walter J. Weber, Jr., Ph.D., P.E., DEE
Earnest Boyce Distinguished Professor
University of Michigan

REVIEWING ORGANIZATIONS

The following organizations contributed to the monograph's review and acceptance by the professional community. The review process employed by each organization is described in its acceptance statement. Individual reviewers are, or are not, listed according to the instructions of each organization.

Air & Waste Management Association

The Air & Waste Management Association is a nonprofit technical and educational organization with more than 14,000 members in more than fifty countries. Founded in 1907, the Association provides a neutral forum where all viewpoints of an environmental management issue (technical, scientific, economic, social, political, and public health) receive equal consideration.

This worldwide network represents many disciplines: physical and social sciences, health and medicine, engineering, law, and management. The Association serves its membership by promoting environmental responsibility and providing technical and managerial leadership in the fields of air and waste management. Dedication to these objectives enables the Association to work towards its goal: a cleaner environment.

Qualified reviewers were recruited from the Waste Group of the Technical Council. It was determined that the monograph is technically sound and publication is endorsed.

The reviewers were:

James R. Donnelly
Director of Environmental Services and Technologies
Davy Environmental

Paul Lear
OHM Remediation Services, Corp.

American Institute of Chemical Engineers

The Environmental Division of the American Institute of Chemical Engineers has enlisted its members to review the monograph. Based on that review the Environmental Division endorses the publication of the monograph.

American Society of Civil Engineers

Qualified reviewers were recruited from the Environmental Engineering Division of ASCE and formed a Subcommittee on WASTECH®. The members of the Subcommittee have reviewed the monograph and have determined that it is acceptable for publication.

Hazardous Waste Action Coalition

The Hazardous Waste Action Coalition (HWAC) is an association dedicated to promoting an understanding of the state of the hazardous waste practice and related business issues. Our member firms are engineering and science firms that employ nearly 75,000 of this country's engineers, scientists, geologists, hydrogeologists, toxicologists, chemists, biologists, and others who solve hazardous waste problems as a professional service. HWAC is pleased to endorse the monograph as technically sound.

The lead reviewer was:

James D. Knauss, Ph.D.
Hatcher-Sayre, Incorporated

Water Environment Federation

The Water Environment Federation is a nonprofit, educational organization composed of member and affiliated associations throughout the world. Since 1928, the Federation has represented water quality specialists including engineers, scientists, government officials, industrial and municipal treatment plant operators, chemists, students, academic and equipment manufacturers, and distributors.

Qualified reviewers were recruited from the Federation's Hazardous Wastes Committee and from the general membership. A list of their names, titles, and business affiliations can be found listed below. It has been determined that the document is technically sound and publication is endorsed.

The reviewers were:

Roger R. Hlavek
Chemical Engineer
Naval Air Warfare Center — Aircraft Division

Murali Kalavapudi
Senior Environmental Engineer
Energetics, Incorporated

Byung R. Kim*
Principal Staff Engineer
Ford Research Laboratory

Edward R. Maziarz
Environmental Engineering Consultant — Industrial Wastes
ALCOA

Charles D. Sweeney
Director
CDS Laboratories, Inc.

Arnold S. Vernick, P.E., DEE
Associate
Geraghty & Miller, Inc.

Scott E. Walters
Environmental Chemist
Pennsylvania Department of Environmental Resources

Robert C. Wichser
Chief Utility Engineer
Richmond Virginia Department of Public Utilities

*WEF lead reviewer

ACKNOWLEDGMENTS

The WASTECH® project was conducted under a cooperative agreement between the American Academy of Environmental Engineers® and the Office of Solid Waste and Emergency Response, U.S. Environmental Protection Agency. The substantial assistance of the staff of the Technology Innovation Office was invaluable.

Financial support was provided by the U.S. Environmental Protection Agency, Department of Defense, Department of Energy, and the American Academy of Environmental Engineers®.

This multiorganization effort involving a large number of diverse professionals and substantial effort in coordinating meetings, facilitating communications, and editing and preparing multiple drafts was made possible by a dedicated staff provided by the American Academy of Environmental Engineers® consisting of:

Paul F. Peters
Assistant Project Manager & Managing Editor

Karen M. Tiemens
Editor

Susan C. Richards
Project Staff Assistant

J. Sammi Olmo
Project Administrative Manager

Yolanda Y. Moulden
Staff Assistant

I. Patricia Violette
Staff Assistant

TABLE OF CONTENTS

LIST OF TABLES

LIST OF FIGURES

1 INTRODUCTION

This monograph on chemical treatment is one of a series of eight on innovative site and waste remediation technologies that are the culmination of a multiorganization effort involving more than 100 experts over a two-year period. It provides the experienced, practicing professional guidance on the application of innovative processes considered ready for full-scale application. Other monographs in this series address bioremediation, soil washing/soil flushing, solvent chemical extraction, stabilization/ solidification, thermal desorption, thermal destruction, and vacuum vapor extraction.

1.1 Chemical Treatment

The term chemical treatment, as used in this monograph, refers to the use of reagents to destroy or chemically modify target contaminants by means other than pyrolysis or combustion. The monograph addresses processes that chemically treat contaminated soils, groundwaters, surface waters, and, to a limited extent, concentrated contaminants. Chemical treatment is a means of converting hazardous constituents into less environmentally objectionable forms in order to meet treatment objectives.

This monograph addresses substitution, oxidation, and chemical precipitation processes. It addresses processes within these classes that are sufficiently advanced for full-scale application. There are a number of emerging technologies within these classes that are in the research or an early development stage, not yet ready for full-scale application, that appear to be very promising technologically. Six such technologies are briefly addressed in Appendix A.

1.2 Development of the Monograph

1.2.1 Background

Acting upon its commitment to develop innovative treatment technologies for the remediation of hazardous waste sites and contaminated soils and groundwater, the U.S. Environmental Protection Agency (EPA) established the Technology Innovation Office (TIO) in the Office of Solid Waste and Emergency Response in March, 1990. The mission assigned TIO was to foster greater use of innovative technologies.

In October of that same year, TIO, in conjunction with the National Advisory Council on Environmental Policy and Technology (NACEPT), convened a workshop for representatives of consulting engineering firms, professional societies, research organizations, and state agencies involved in remediation. The workshop focused on defining the barriers that were impeding the application of innovative technologies in site remediation projects. One of the major impediments identified was the lack of reliable data on the performance, design parameters, and costs of innovative processes.

The need for reliable information led TIO to approach the American Academy of Environmental Engineers®. The Academy is a long-standing, multidisciplinary environmental engineering professional society with wide-ranging affiliations with the remediation and waste treatment professional communities. By June 1991, an agreement in principle (later formalized as a Cooperative Agreement) was reached. The Academy would manage a project to develop monographs describing the state of available innovative remediation technologies. Financial support would be provided by the EPA, U.S. Department of Defense (DOD), U.S. Department of Energy (DOE), and the Academy. The goal of both TIO and the Academy was to develop monographs providing reliable data that would be broadly recognized and accepted by the professional community, thereby eliminating or, at least, minimizing this impediment to the use of innovative technologies.

The Academy's strategy for achieving the goal was founded on a multiorganization effort, WASTECH® (pronounced Waste Tech), which joined in partnership the Air and Waste Management Association, the American Institute of Chemical Engineers, the American Society of Civil Engineers, the American Society of Mechanical Engineers, the Hazardous

Waste Action Coalition, the Society for Industrial Microbiology, and the Water Environment Federation, together with the Academy, EPA, DOD, and DOE. A Steering Committee composed of highly respected representatives of these organizations having expertise in remediation technology formulated the specific project objectives and process for developing the monographs (See page iv for a listing of Steering Committee members).

By the end of 1991, the Steering Committee had organized the Project. Preparation of the monograph began in earnest in January, 1992.

1.2.2 Process

The Steering Committee decided upon the technologies, or technological areas, to be covered by each monograph, the monographs' general scope, and the process for their development and appointed a task group composed of five or more experts to write a manuscript for each monograph. The task groups were appointed with a view to balancing the interests of the groups principally concerned with the application of innovative site and waste remediation technologies — industry, consulting engineers, research, academe, and government (see page iii for a listing of members of the Chemical Treatment Task Group).

The Steering Committee called upon the task groups to examine and analyze all pertinent information available, within the Project's financial and time constraints. This included, but was not limited to, the comprehensive data on remediation technologies compiled by EPA, the store of information possessed by the task groups' members, that of other experts willing to voluntarily contribute their knowledge, and information supplied by process vendors.

To develop broad, consensus-based monographs, the Steering Committee prescribed a twofold peer review of the first drafts. One review was conducted by the Steering Committee itself, employing panels consisting of two members of the Committee supplemented by at least four other experts (See *Reviewers,* page iii, for the panel that reviewed this monograph). Simultaneous with the Steering Committee's review, each of the professional and technical organizations represented in the Project reviewed those monographs addressing technologies in which it has substantial interest and competence. Aided by a Symposium sponsored by the Academy in October, 1992, persons having interest in the technologies were encouraged to participate in the organizations' review.

Comments resulting from both reviews were considered by the Task Group, appropriate adjustments were made, and a second draft published. The second draft was accepted by the Steering Committee and participating organizations. The statements of the organizations that formally reviewed this monograph are presented under *Reviewing Organizations* on page v.

1.3 Purpose

The purpose of this monograph is to further the use of innovative chemical treatment site remediation and waste processing technologies, that is, technologies not commonly applied, where their use can provide better, more cost-effective performance than conventional methods. To this end, the monograph documents the current state of a number of innovative chemical treatment technologies.

1.4 Objectives

The monograph's principal objective is to furnish guidance for experienced, practicing professionals and users' project managers. The monograph is intended, therefore, not to be prescriptive, but supportive. It is intended to aid experienced professionals in applying their judgment in deciding whether and how to apply the technologies addressed under the particular circumstances confronted.

In addition, the monograph is intended to inform regulatory agency personnel and the public about the conditions under which the processes it addresses are potentially applicable.

1.5 Scope

The monograph addresses innovative chemical treatment technologies that have been sufficiently developed so that they can be used in full-scale applications. It addresses all aspects of the technologies for which sufficient data were available to the Chemical Treatment Task Group to describe and explain the technologies and assess their effectiveness, limitations, and potential applications. Laboratory- and pilot-scale studies were addressed, as appropriate.

The monograph's primary focus is site remediation and waste treatment. To the extent the information provided can also be applied to production waste streams, it will provide the profession and users this additional benefit. The monograph considers all waste matrices to which chemical treatment processes can be reasonably applied, such as soils, liquids, and sludges.

Application of site remediation and waste treatment technology is site specific and involves consideration of a number of matters besides alternative technologies. Among them are the following that are addressed only to the extent essential to understand the applications and limitations of the technologies described:

- site investigations and assessments;
- planning, management, specifications, and procurement;
- regulatory requirements; and
- community acceptance of the technology.

1.6 Limitations

The information presented in this monograph has been prepared in accordance with generally recognized engineering principles and practices and is for general information only. This information should not be used without first securing competent advice with respect to its suitability for any general or specific application.

Readers are cautioned that the information presented is that which was generally available during the period when the monograph was prepared. Development of innovative site remediation and waste treatment technologies is ongoing. Accordingly, postpublication information may amplify, alter, or render obsolete the information about the processes addressed.

This monograph is not intended to be and should not be construed as a standard of any of the organizations associated with the WASTECH® Project; nor does reference in this publication to any specific method, product, process, or service constitute or imply an endorsement, recommendation, or warranty thereof.

1.7 Organization

This monograph and others in the series are organized under a uniform outline intended to facilitate cross reference among them and comparison of the technologies they address. Chapter 2.0, Process Summary, provides an overview of all material presented. Chapter 3.0, Process Identification, provides comprehensive information on the processes addressed. Each process is fully analyzed in turn. The analysis includes a description of the process (what it does and how it does it), its scientific basis, status of development, environmental effects, pre- and posttreatment requirements, health and safety considerations, design data, operational considerations, and comparative cost data to the extent available. Also addressed are process-unique planning and management requirements, and process variations.

Chapter 4.0, Potential Applications, Chapter 5.0, Process Evaluation, and Chapter 6.0, Limitations, provide a synthesis of available information and informed judgments on the processes. Each of these chapters addresses the processes in the same order as they are described in Chapter 3.0. Chapter 7.0, Technology Prognosis, identifies likely future applications of the processes.

2

PROCESS SUMMARY[1]

Chemical treatment, for the purposes of this monograph, refers to the use of reagents to destroy or chemically modify contaminants by means other than pyrolysis, wet oxidation, or combustion. This monograph addresses techniques used to chemically treat contaminated soils, groundwaters, surface waters, and, to a limited extent, concentrated contaminants.

These technologies embrace a wide variety of processes used in treating wastes and contaminated materials. They consist of a series of techniques that can be selectively applied to destroy or modify organic and inorganic contaminants. The selection varies depending on the particular contaminants and media. The systems and processes usually have to be modified from one site to the next. Chemical treatment is usually employed as a pre- or posttreatment process in site remediation and seldom as a stand-alone process.

Chemical treatment processes convert hazardous constituents into less objectionable environmental forms in order to meet treatment objectives. The ideal goal for a treatment process is the complete mineralization of the target contaminants, for example, the reduction of polychlorinated biphenyls (PCBs) to sodium chloride, carbon dioxide, and water. This outcome is relatively rare, however, and the goal set for most chemical treatment processes is more modest — the conversion of selected contaminants into less toxic or unregulated, i.e., not governed by state or federal regulations, chemical forms. In many cases, the long-term environmental effect of the chemical reagents and reaction products may not be well understood, and it may not be known whether they will remain stable for long periods. This uncertainty leads to tradeoffs between present cost and the risk of future adverse environmental impact. The toxicity of the reaction products of all chemical treatment methods, especially those of substitution reactions,

1. This chapter is a summary of Chapters 3.0 through 7.0. Sources are cited, where appropriate, in those chapters — Ed.

needs to be addressed. For example, replacing chlorine atoms on a contaminant with a glycol structure of methoxyethoxy moiety (as in one of polyethylene glycol processes described in the text) may produce a molecule that may not be regulated at the time of treatment, but it may not be less toxic than the original compound.

2.1 *Process Identification and Description*

Although chemical treatment processes potentially span the full range of chemistry, only the following techniques have been applied in site and waste remediation:

- Substitution;
- Oxidation; and
- Precipitation.

See table 2.1 for a list of the contaminants, by media, that might be treated by each technique.

2.1.1 Substitution Processes

Substitution reactions have been used to treat soils contaminated with chlorinated organic compounds, such as PCBs or chlorodibenzodioxins.

Table 2.1
Applicability of Chemical Treatment Techniques

Technique	Media	Contaminants
Substitution Processes	Soil, oils, and debris Generally not applicable to aqueous streams	Halogenated and other substituted organics, such as PCB and dioxins
Oxidation Processes	Aqueous streams, dilute particulate suspensions, and some debris	Relatively low concentrations of organics
Precipitation Processes	Only aqueous streams	Inorganic contaminants

Substitution reactions do not mineralize the halogenated organic compounds, but, instead, convert the compounds into substituted forms that are within regulatory standards. In most cases, the chlorine or other group that is characteristic of the target compound is substituted by another functional group that converts the contaminant to a nonregulated form.

The substitution processes that have been used or appear to be developed sufficiently to be used for site and waste remediation fall into two broad classes: (1) low-temperature processes in the 130 to 160°C (260 to 320°F) range and (2) high-temperature processes in the 200 to 345°C (400 to 650°F) range.

Three innovative low-temperature substitution processes were identified and are addressed in this monograph — the potassium polyethylene glycol (KPEG) Process, the Galson Research Corporation (GRC) Process, and the KGME/DECHLOR Process. Of these, only the GRC Process has been used commercially. The KPEG Process was used once at the pilot-scale. The KGME/DECHLOR Process is, at present, in the pilot-scale phase of development and it appears that it will be applicable to site and waste remediation.

One apparently successful high-temperature process (SoilTech Anaerobic Thermal Processor (ATP)), which combines chemical treatment with pyrolysis, and is addressed in this monograph. Its reactor subjects the treated material first to temperatures of 200° to 340°C (400 to 650°F) and then to temperatures of 480 to 600°C (900 to 1,100°F).

One emerging substitution reaction technology, Base Catalyzed Decomposition (BCD), has been identified. It, along with other emerging technologies, is discussed in Appendix A.

2.1.2 Oxidation Processes

In the oxidative degradation of organic compounds, an organic compound is converted by means of an oxidizing agent into new materials typically having either a higher oxygen or lower hydrogen content than the original compound. Oxidative processes that appear applicable use ozone and hydrogen peroxide (individually or together) in conjunction with ultraviolet light to destroy organic contaminants in an aqueous stream.

The oxidation processes addressed in this monograph are ozone-based advanced oxidative processes (AOPs). These processes, based on free radical, chain reaction chemistry, combine ozonation with ultraviolet (UV) photolysis. Some compounds that are resistant to destruction or are only slowly destroyed by UV irradiation alone or UV in combination with either ozone or hydrogen peroxide are rapidly destroyed, and to a greater extent, by all three agents. Following are the AOPs addressed:

- Rayox — This process has been shown to be effective in treating a wide variety of halogenated compounds, volatile compounds, and other organics; and
- Ultrox — This process has been used in treating a variety of organic constituents in aqueous streams.

The following emerging oxidation technologies are addressed in appendix A:

- Iron (II) Catalyzed H_2O_2 Oxidation (Fenton's Reagent);
- Photocatalysis in Semiconductor Systems;
- Ionizing Radiation;
- Sonication; and
- Iron (VI) Oxidation.

A case study of the use of hypochlorite in treating cyanide-contaminated material is provided in Appendix B. This process is well established and of relatively limited applicability, but the case study is presented for the useful information it provides on the cost of treatment.

2.1.3 Chemical Precipitation Processes

Chemical precipitation entails transforming a soluble compound into an insoluble form through the addition of chemicals, to a point of supersaturation. The process is routinely used in treating wastewaters. Its application in site remediation is less common, but it appears to have potential for removing toxic metals.

Precipitation is largely used for the treatment of aqueous materials contaminated with toxic inorganic elements and compounds. In treating soils, it would normally be classified as a stabilization rather than a chemical treatment process.

The following precipitation processes are addressed:

- Hydroxide;
- Carbonate;
- Sulfide;
- Xanthate; and
- Combined precipitation treatment.

2.1.4 Design and Other Considerations

Many of the general design considerations for chemical treatment processes are similar to those of other remediation processes. Design considerations unique to chemical treatment are addressed in this monograph. See table 2.2 (on page 2.6) for a summary listing of some unique design considerations along with health and safety considerations and pre- and posttreatment requirements.

2.1.5 Materials Handling

Once the fundamental requirement of performance needs is met, the principal concern in selecting a chemical treatment system is meeting material handling requirements. The system must accommodate wastes, often abrasive and corrosive, of varying characteristics. In many cases, the wastes may be incompatible with particular designs. For example, high concentrations of suspended solids may make it impossible to use UV-based processes because turbidity will attenuate UV transmission through the suspension.

Fugitive emissions are an important consideration. Conveyors and other materials-moving equipment must often be modified to minimize fugitive emissions. A conveyor that releases a small amount of dust may be acceptable for transporting gravel, but not for transporting a contaminated material.

Another important design consideration lies in recognizing that most wastes are heterogeneous. The system must be capable of treating extremes in waste composition and contaminant concentrations. If, for example, the system is designed to handle wastes having average characteristics, it will fail when it is fed waste having characteristics outside this limited range.

Table 2.2

Summary Listing of Major Design and Other Considerations

	Design Considerations	Requirements	Posttreatment Requirements	Health & Safety Considerations
Substitution Processes	▪ Highly alkaline reagents will attack aluminum and magnesium system components, forms H_2 gas ▪ Elastomeric seals are subject to attack by reagents ▪ Reactor size depends on residence time required	▪ Conduct treatability studies ▪ Dewater and filter oils ▪ Delump and screen soils ▪ Drain and dry solids and soils	▪ Neutralize excess caustic ▪ Remove excess reagents by washing	▪ Uncertainty about long-term effects of substitution products ▪ Avoid aluminum and magnesium for material of construction
Oxidation Processes	▪ System design depends on residence time requirements and level of degradation needed ▪ Performance depends on UV lamp geometry, wavelength of UV, and optical path length ▪ Efficiency of ozone generation depends on type and pretreatment of feed gas–air or oxygen	▪ Filter liquid streams ▪ Pretreatment may be needed to prevent excessive fouling of UV lamps	▪ Residual ozone in offgas must be destroyed ▪ Effluent may require pH adjustment, solids removal, or removal of residual H_2O_2 ▪ Effluent may require further treatment, i.e., biotreatment or adsorption	▪ Handling of strong oxidizing agents ▪ Control of hydrogen gas ▪ Ozone is a toxic gas and a fire hazard
Precipitation Processes	▪ Solubility products of target species and precipitating reagent must be below their concentrations in solution ▪ If multiple substances are present, coprecipitation can occur	▪ Filter liquid streams ▪ pH adjustment	▪ Highly site specific ▪ Must consider sludge volume and stability ▪ Cost savings may result from metals recovery	▪ Potential for H_2S formation in sulfide precipitation

Heterogeneity of wastes presents a problem for all remediation processes, but chemical treatment processes are particularly vulnerable because of competing reactions, stoichiometric relationships among reactants, and the possibility that trace materials can poison reactions, especially catalytic or chain reactions.

2.1.6 Costs

Major components of fixed cost are equipment amortization, marketing, permits and approvals, and shipping and setting up equipment. Major components of variable cost are reagent purchase and recovery, utilities, labor, travel and subsistence, and analytical costs.

Most remediation systems tend to be highly specialized. The system is designed for a given application and its cost must be amortized over one or a few projects. One way to minimize this effect is to design the system so that it can be assembled from readily available, reusable components. While the system may be highly specialized, the components can be used from one project to the next. Chemical treatment systems lend themselves to this approach. The components are generally readily available and can be rearranged for different applications. In addition, the systems tend to be compact. They are relatively inexpensive to ship, set up, and knock down.

Reagent cost, unique to chemical treatment processes, is usually a significant part of the overall cost, and reagent recovery is an economic (as well as environmental) necessity. In addition, the typical chemical treatment process requires highly trained workers, usually unavailable locally. Consequently, the labor and travel costs for a chemical treatment remediation are generally higher than those of other remediation operations.

2.2 *Potential Applications*

Chemical treatment is rarely used alone. Typically, it is used as part of a treatment train — as a pretreatment technique to enhance the efficiency of subsequent processes or in posttreatment of an effluent. Following are examples of applications:

- Oxidation techniques have been used in wastewater treatment to "soften" organic compounds to improve their biodegradability; and
- Chemical dechlorination can be used to treat the chlorinated organics that are removed from soil through solvent extraction or thermal desorption.

The waste matrix is a major factor in any remediation. With soils, such variables as the quantity and quality of the natural organic material, mineral content, and amorphousness of clays result in a high variability of performance among sites as well as within a particular site. Designs based on "average characteristics of a soil" will often encounter operating difficulties when a "nonaverage" soil is encountered. Nevertheless, the following general rules appear to apply, although with numerous site- and process-specific exceptions:

- Solids tend to be more difficult to treat than liquids;
- The larger the particle size of nonporous materials in the matrix, the easier it is to treat the contaminants. Larger particulates make the contaminants more accessible to the reagent and permit a clean separation of the reagent after treatment (the opposite tends to be true of porous materials, such as organic soils); and
- The cost of treatment of substitution processes increases with contaminant concentration. For oxidation processes, the relationship is less clear. Precipitation processes typically do not apply to low-concentration contaminants.

Chemical treatment should be considered for use at sites where one or more of the following conditions exist:

- The remediation is mainly driven by one or a few specific kinds of contaminants that can be chemically modified;
- The quantities of material to be treated or local concerns preclude the transport of the contaminants to an off-site treatment location;
- Established processes, such as incineration, are unacceptable technically or because of local concerns; and
- The quantity of material to be treated is small or the contaminant concentrations are low. The economics of such a site favor the low-capital, high-operating cost approach.

2.2.1 Substitution Processes

Substitution processes should be considered for treating oils or soils and other solids that are contaminated with a particular class of compound, such as PCBs. They do not appear to be effective in treating aqueous streams. Substitution processes are most effective in treating halogenated aromatic compounds. Possible additional applications include treating halogenated aliphatics, nitrogen-bearing compounds, and sulfonated compounds.

2.2.2 Oxidation Processes

Oxidation processes should be considered for treating aqueous streams and some slurries that are considered hazardous because they contain low concentrations of organic constituents. The most likely applications are treatment of groundwater or surface water streams that are contaminated with such compounds as cyanides and light organics.

2.2.3 Precipitation Processes

Precipitation processes, by their nature, are limited to liquid systems. They should be considered for treating materials that are hazardous because they contain toxic metal compounds in an aqueous solution. Precipitation is applicable whenever the target metal is present in a soluble form and can be chemically converted to a less soluble form.

2.3 Process Evaluation

Both oxidation and precipitation processes are commercially available for treating drinking water and wastewater. Therefore, it is believed these processes would be effective in treating contaminated groundwaters. Their use for this purpose is a matter of transfer of the technology to another application, rather than *de novo* development of technology .

For over a decade, substitution processes have been available for the treatment of soils and sludges contaminated with PCBs and other chlorinated organics. They have not been used extensively for the following reasons:

- Existing technologies, such as incineration and landfilling have been cheaper and more widely available; and
- An initial field test of the technology identified design problems, but no additional funding for follow-up work was available.

It appears these problems are being resolved or eliminated. Incineration and landfilling are, at times, rejected. In addition, the costs quoted by the vendors of several substitution processes (GRC, SoilTech/ATP, and KGME/DECHLOR), are beginning to approach those of incineration. Finally, the design problems are being addressed and solved.

2.4 *Limitations*

2.4.1 Substitution Processes

Although substitution reactions can occur in the presence of water, large quantities of water will interfere with the desired chemical reactions and consume excessive quantities of reagent. Therefore, substitution processes are not now practical for treating aqueous wastes. Although the processes can operate in the presence of water, the processes are generally impractical because their operating temperatures above 100°C boil the water out of the reaction system, increase heat requirements, and reduce efficiency. They may be operated under pressure in order to maintain a temperature above the system's boiling point. However this requires special pressure vessels and other process equipment.

Commercial substitution reactions presently available do not mineralize halogenated organic compounds. Instead, they convert the compounds into substituted forms that are within regulatory standards. This means that they are applicable only in treating regulated contaminants, such as PCBs and dioxins. They are not applicable in treating general types of contaminants, such as hydrocarbons or unsubstituted organic compounds, such as benzene.

2.4.2 Oxidation Processes

Of the three classes of processes addressed here, oxidation processes actually destroy or mineralize the target organic contaminants. In doing so, however, they consume relatively large quantities of ozone and/or hydrogen peroxide.

Advanced oxidation processes using ultraviolet light to form free radicals were developed to overcome this limitation. Since advanced oxidation processes are based on hydroxyl free radical chemistry, chemical interactions are highly nonspecific and nonselective. Rates of destruction vary with such factors as the nature of the contaminant mixture, pH, concentration of contaminants, presence of scavengers, and inorganic nature.

Oxidation processes do not work well in the presence of free radical scavengers, such as bicarbonate and carbonate ions. The scavengers consume the ozone and hydrogen peroxide, and inhibit the effect of the UV radiation. The presence of such scavengers requires higher doses of oxidizers and larger UV fluxes.

Another important factor is penetration of UV light through the wastewater stream. Light penetration is attenuated by high particle concentrations. Consequently, the technique, in general, is not well suited to treating soils. A similar problem is optical fouling of the quartz tubes containing the UV light source. This occurs gradually, over the course of the treatment, and can significantly reduce process performance and efficiency. Some form of tube cleaning should be incorporated in all such processes.

2.4.3 Precipitation Processes

Precipitation reactions are limited to the treatment of inorganic materials in aqueous media. They are routinely used in the treatment of wastewaters, but their application in site remediation is less common.

One potential limitation lies in the collection and handling of precipitates. Their handling must be addressed in a treatability study conducted before the treatment method is selected.

2.5 Technology Prognosis

Under proper conditions, discussed in this monograph, chemical treatment can be a useful treatment technology. The following are likely applications of the processes addressed in this monograph:

- Substitution processes, especially the high temperature substitution processes, will be used to treat soils and sludges contaminated with PCBs, pentachlorophenols, chlorodibenzodioxins, and chlorodibenzofurans;

- Oxidation and precipitation processes will be used to treat water from pump-and-treat applications; and
- Precipitation processes will be commonly used to treat aqueous streams that are contaminated with toxic metals.

3

PROCESS IDENTIFICATION AND DESCRIPTION

Chemical treatment, for the purposes of this monograph, refers to the use of reagents to destroy or chemically modify target contaminants by means other than pyrolysis or combustion. The monograph addresses techniques used to chemically treat contaminated soils, groundwaters, surface waters, and, to a limited extent, concentrated contaminants. Chemical treatment is a means of converting hazardous constituents into less environmentally objectionable forms in order to meet treatment objectives.

Embracing a wide variety of processes used in treating wastes and contaminated materials, chemical treatment is not a "technology" in the same sense as is, for example, incineration. Instead, it consists of a series of techniques that can be selectively applied to destroy or modify organic and inorganic contaminants. The selection varies depending on the particular contaminants and media. The systems and processes usually have to be modified, at least to some extent, from one site to the next.

Chemical treatment has been successful in a number of applications. The following instances are limited to those for which data exist from (1) an actual remediation, (2) a scale equivalent to a remediation, or (3) a pilot study whose results can be scaled up to a remediation application:

- treatment of polychlorinated biphenyls (PCBs), chlorodibenzodioxins (dioxin) and pentachlorophenol (PCP) in soil;
- treatment of cyanides in water and on debris;
- treatment of phenols in groundwater;
- treatment of nonsubstituted and chlorinated hydrocarbons in soil and in groundwater;
- precipitation of toxic metals from groundwater;

- treatment of PCBs in mineral oil; and
- destruction of low-level organic compounds in groundwater.

Chemical treatment is rarely used as the sole process. It may be used as a pretreatment technique to enhance the efficiency of subsequent processes or as a posttreatment step in treating an effluent. Following are examples of such applications:

- Various advanced oxidation techniques have been successfully employed to soften organic compounds to improve their biodegradability;
- Chemical dechlorination can be used to treat the contaminated eluate from solvent extraction of chlorinated organics from soil; and
- Chemical destruction can be used to treat the offgases from a vapor phase extraction process.

The ideal goal for a treatment process is the complete mineralization of the contaminants — for example, reduction of a PCB to sodium chloride, carbon dioxide and water. Although mineralization can result from certain oxidation processes, this outcome is relatively rare and has been commercially achieved with but a few readily oxidizable compounds. Therefore, the actual goal set for chemical treatment processes is more modest — for example, the conversion of selected contaminants into chemical forms that are less toxic or unregulated.

In many cases, the long-term stability or environmental effect of the chemical reagents and the reaction products may not be well understood and it may not be known whether they will remain stable for long periods. This uncertainty results in tradeoffs between present costs and the risk of future liability or future adverse environmental impact. As considered here, chemical treatment has the following specific goals:

- Convert the hazardous constituents into a less toxic or environmentally less objectionable form. For example, the replacement of chlorine on a PCB or chlorodibenzodioxins (dioxin) molecule with an aryl or alkyl group (using, for example, sodium napthalenide reagent) or with another functional group (as with a polyethylene glycol). The addition of the group to the PCB or dioxin molecule converts it to an unregulated substance;

- Convert the hazardous constituents into a less mobile form, for example, by precipitation;
- Convert the hazardous constituent into a form that is more amenable to subsequent treatment by another process. An example is the partial oxidation of contaminants in groundwater to convert refractory (difficult to degrade) organics into compounds that are amenable to biodegradation; and
- Convert the hazardous constituent into a more mobile form, thereby making it amenable to a second treatment process that is to remove the modified hazardous constituents from the nonhazardous matrix. (While mobilization is conceptually possible, no such commercial or near-commercial chemical processes were identified).

An important consideration when using chemical treatment is the nature of the material leaving the treatment process. Chemical treatment requires that reagents be mixed with the contaminated material. When the reagents destroy or modify the target contaminants (e.g., Cr^{+6}, PCBs, dioxins), the "decontaminated material" still contains chemical reaction products and any residual reagents, which may be toxic or hazardous. Even if they are not, their presence may have a significant environmental impact, such as a high-oxygen demand, on the surrounding ecosystem. If the "treated" material is returned to the site, the residual reagents (which will usually be mobile) may be cause for concern. It might not be possible to dispose of the treated material because of health and safety regulations.

An example of this concern arises in the treatment of organic contaminants by substitution processes. As explained below, all commercial substitution processes convert the contaminant into an unregulated form by replacing one or more halogen atoms on the target molecule (a PCB, for example) with another functional group, such as an ether. The resultant compound is, at present, unregulated, but its environmental impact still needs to be considered. Complete replacement with hydrogen (as claimed for the Base Catalyzed Decomposition (BCD) process) may not always be desirable. For example, if all chlorine atoms on chlorobenzene are replaced with hydrogen, benezene, a known carcinogen, is formed. One must approach chemical treatment with knowledge of the chemical transformations, regulatory requirements, and environmental ramifications.

Chemical treatment, by its nature, is technique-oriented rather than process-oriented. That is, to determine the proper treatment method, it is first necessary to identify the target contaminant and determine its reactivity and accessibility. Chemical knowledge is used to ascertain the kind of chemical reactions to which the target compound(s) is amenable, evaluate the available equipment, and select or design the appropriate treatment system. This orientation is reflected in the grouping of technologies addressed in this monograph by the kind of chemical reactions they use, that is, by substitution processes, oxidation processes, and precipitation processes (see table 3.1 on page 3.5).

While substitution reactions can occur in the presence of water, large quantities of water will interfere with the desired chemical reactions and consume excessive quantities of reagent. As a result, these reactions have not, to date, been applied to the treatment of aqueous systems.

Oxidation embraces a broad class of chemical reactions that are well established methods for treating aqueous liquids containing small amounts of organics or cyanide. One can effectively oxidize low concentrations of contaminants by appropriate use of oxidation reactions; some materials, such as cyanides, can even be mineralized. Commonly used oxidizing agents are ozone (O_3), hypochlorite, and hydrogen peroxide (H_2O_2). Air is frequently used as the oxidizing agent in wet-air oxidation and incineration processes. Reactions using ozone, hydrogen peroxide, ultraviolet (UV) light, and hypochlorite are oxidation processes.

Chemical precipitation involves transforming a soluble compound into an insoluble form through the addition of chemicals such that a supersaturated environment exists. Precipitation is largely used for the treatment of aqueous materials contaminated with toxic inorganic elements and compounds. Its use in the treatment of soils would normally be considered to be stabilization. The process has been routinely used for the treatment of wastewaters. Its application in remediation is less common, but data from wastewater applications indicate that it is applicable. No specific use of the process was found that would qualify it as an "innovative technology" (see Section 1.3). Its underlying principles are presented here in the interest of aiding in transferring the technology from the field of wastewater treatment to the treatment of groundwater, surface waters, and other liquids in remediations.

Table 3.1

Chemical Treatment Processes Addressed

•SUBSTITUTION PROCESSES

ESTABLISHED TECHNOLOGIES

- Sodium metal/aromatic/ether
- NaPEG* - developed in early 1980s — no successful field tests conducted

INNOVATIVE TECHNOLOGIES

Low Temperature PEG reagents

- KPEG/APEG
    - APEG Plus - Galson Research (GRC)
- DECHLOR/KGME - Chemical Waste Management (CWM)

High Temperature Chemical Reactions

- Soiltech Anaerobic Thermal Processor (ATP) Canonie Engineering

EMERGING TECHNOLOGY

High Temperature Chemical Reactions

- Base Catalyzed Dechlorination (BCD) - EPA/RREL

•OXIDATION PROCESSES

INNOVATIVE TECHNOLOGIES

- Ozonation
- H_2O_2 or Ozone with UV

EMERGING TECHNOLOGIES

- Iron (II) Catalyzed Oxidation, (H_2O_2/FeII), Fenton's Reagent
- Sonication
- Ionizing Radiation

•CHEMICAL PRECIPITATION PROCESSES

ESTABLISHED TECHNOLOGIES

- Hydroxide Precipitation
- Carbonate Precipitation
- Sulfide Precipitation
- Xanthate Precipitation
- Combined Precipitation Treatment

* The NaPEG process is discussed as a precursor process which, while not commercially sucessful on its own, illustrates the technical basis for other processes which were operated commercially.

An established method of chemical treatment is the use of sodium or calcium hypochlorite (NaOCl or $Ca(OCl)_2$) for disinfection, by destroying low concentrations of organics in water and for the destruction of cyanides by oxidation. A discussion of the treatment of cyanide-contaminated film chips as part of the emergency removal action at the PBM Enterprises site in Romulus, Michigan is included in Appendix B. This application is not considered an innovative process; destruction of cyanides by hypochlorite is established technology and chemistry. The case study, however, illustrates when chemical treatment is applicable and provides a concrete example of the costs of such an application.

For operational reasons, not every technology can be applied to media encountered in remediation — soil, debris, water, and oil. See table 3.2.

Table 3.2
Contaminants and Media that have been Chemically Treated

SOILS
- PCB, dioxins, PCP

DEBRIS
- cyanides

WATER
- cyanides
- phenols
- metals (precipitation)

OTHER
- PCB and dioxins in mineral oil

3.1 Substitution Processes

Commercial processes based on substitution reactions have successfully treated soils, sludges, and oils contaminated with halogenated organics, such as PCBs, chlorodibenzodioxins (dioxins), or chlorodibenzofurans (dibenzofurans). It is claimed that an emerging process, the BCD Process, discussed in Appendix A, is capable of also treating hazardous sulfur-bearing and nitrogen-bearing compounds. However, only laboratory data on this application are available at present. Substitution reactions have been used successfully in remediations (up to 36,000 tonne (40,000 ton)) to treat soils contaminated with chlorinated organic compounds. Examples of treatment are:

- Conversion of PCBs on soil and in oils to unregulated, substituted chlorobiphenyls; and

- Conversions of chlorodibenzodioxins and chlordibenzofurans on soil and in oil to substituted chlorodibenzodioxins and chlordibenzofurans which may be subject to less stringent regulation.

Of the two substitution processes listed in table 3.1 (on page 3.5) under "established technology," the first, Sodium metal/aromatic/ether, is a highly specialized process used commercially in the early to mid-1980s to treat PCB-contaminated mineral oils. The process destroys the PCBs dissolved in the mineral oil by using a reagent prepared by reacting sodium metal, a polycyclic aromatic compound, and an ether. The commonly used aromatic compounds were naphthalene or biphenyl. The commonly used ethers were tetrahydrofuran or one of the diethers, such as ethyleneglycol dimethyl ether (commercial name — Diglyme). The resulting reagent is deactivated by water. Its use, therefore, is limited to high purity organic media.

The second of the established substitution processes, sodium polyethylene glycol (NaPEG), was developed in the late 1970s but was not used beyond the laboratory scale, except for a few unsuccessful field trials. Sodium polyethylene glycol is formed by reacting sodium hydroxide (NaOH) with polyethylene glycol. The resultant reagent is mixed with the contaminated material and the NaPEG, in principle, replaces one or more chlorines on the target molecule.

Subsequent research indicated that a similar substitution reaction using potassium polyethylene glycol (KPEG) is preferable to the NaPEG (Brunelle and Singleton 1983, 1985). The KPEG reagent is prepared in the same way as the NaPEG reagent except that potassium hydroxide (KOH) is used instead of sodium hydroxide. The processes are described in greater detail by Weitzman (1982), Brown et al. (1982), and Smith and Bubbar (1979).

All substitution processes reviewed appear to fall into two broad categories:

- Low temperature processes operating in the 130 to 160°C (260 to 320°F) range; and
- High temperature processes operating in the 200 to 345°C (400 to 650°F) range.

One high temperature process (SoilTech Anaerobic Thermal Processor (ATP)), discussed later, combines chemical treatment with pyrolysis. In the reactor, the treated material is subjected to temperatures of 200 to 345°C (400 to 650°F) in the first zone and then to temperatures of 480 to 600°C (900 to 1,100°F) in the second zone.

Figure 3.1 illustrates the chemistry of substitution reactions. It shows a PCB molecule before and after treatment with (in this case) polyethylene glycol reagent. As can be seen, treatment converts the PCB molecule into a unregulated substituted chlorobiphenyl. Figure 3.1 shows only one level of substitution. Clearly, the chemical reaction can substitute additional chlorine atoms on the PCB molecule. The type of functional group that replaces the chlorine atom is determined by the process chemistry.

Figure 3.1

Chemical "Dechlorination" of PCB

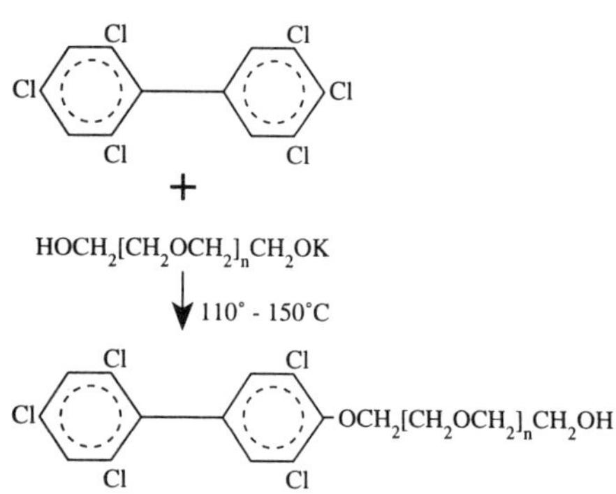

Reprinted by permission of the National Ground Water Association from "Groundwater Treatment With Ultraviolet Light and Hydrogen Peroxide" by M.A. Rowland in the 1989 Outdoor Action Conference. Copyright 1989 by the National Ground Water Association.

Different chemical processes substitute different functional groups for the chlorine. For example, the processes using polyethylene glycol (KPEG and Galson Research Corporation (GRC)) replace the chlorine with the glycol structure; it is claimed that the BCD Process replaces it with hydrogen. Note that data documenting this result have not, to date, been obtained from the developer of the BCD Process .

The main application of the substitution processes has been in the treatment of oils and soils contaminated with halogenated aromatics, mainly PCBs and PCP. The PCBs in the environment are always associated with chlorobenzenes, which have also been treated successfully by substitution processes.

No field data was found on the treatment of halogenated organic compounds other than on the treatment of chlorinated aromatic compounds. Informal discussions with chemists who have worked with substitution reactions led to the impression that:

- Aliphatic compounds appear to be more difficult to treat than aromatic compounds;
- Fluorinated compounds appear to be more difficult to treat than their corresponding chlorinated compounds; and
- Brominated and iodine compounds appear to be less difficult to treat than their corresponding chlorinated compounds.

Although no actual data substantiating these claims was found, this observation may be useful for preliminary evaluations of various processes.

3.1.1 Low-Temperature Substitution Processes

Three low-temperature substitution innovative processes applicable in remediation were identified: KPEG, GRC, and KGME/DECHLOR Processes. The core of each process is a heated, agitated reactor. Heat is necessary to maintain the 130 to 160°C (260 to 320°F) temperatures needed for the reaction. In the KPEG Process, heat is also necessary to lower the viscosity of the polyethylene glycol (PEG)-400 reagent, a viscous fluid that is difficult to mix into the soil. In the KPEG and GRC Processes, the reactor is capable of accepting and discharging soils. The KGME/DECHLOR Process is designed to treat the contaminated oily extract from a thermal treatment process (Chemical Waste Management's X*TRAX Process), and not to treat soil.

The KPEG Process was used only for a demonstration of the chemistry. For this demonstration, the treated soil leaving the reactor did not require treatment beyond neutralization of the reagent to achieve an acceptable pH. The PEG and products of the reaction remained with the soil. As a result, this process did not require equipment to clean the soil. Polyethylene glycol is a food additive and is biodegradable. Its subsequent biodegradation in the environment must, however, be monitored to assure that excessive oxygen demand is not placed on the ecosystem. The GRC Process has an extensive system for washing the residual reagent and reaction products from the soil. The washing serves two purposes: (1) mitigating the environmental impacts discussed above and (2) reducing cost by recovering the PEG. Polyethylene glycol is an expensive material, costing approximately \$2.25/kg (\$1.00/lb), and its recovery and reuse is crucial to the economic viability of the process. See table 3.3 for the extent of destruction of PCBs achieved by the process.

Table 3.3
Results of KPEG Process Guam Application

Sample	PCB Concentration in Soil (ppm)			PCB in Condensate (ppm)
	Before treatment	After treatment	% Reduction	
1	3,276	13.9	99.57	0.831
2	3,828	1.01	99.97	0.176
3	3,651	3.31	99.91	3.50

3.1.1.1 KPEG Process

The process shown in figure 3.2 (on page 3.11) was a batch process employing a combination of PEG-400 (polyethylene glycol with an approximate molecular weight of 400) and KOH to react with the aromatic chlorides in contaminated soils and sludges. Contaminated solids were mixed with the solution of KOH in polyethylene glycol and heated to 150 to 180°C (300 to 360°F) for four hours. This resulted in deactivation of PCBs; that is, one or more of the chloride moieties on the PCBs were replaced by alkoxy ethylene units. At the conclusion of the operation the reaction mixture was highly alkaline. After cooling, the mixture was neutralized by the addition of sulfuric acid and was discharged to a collection hopper made of steel. It was stored until approval was obtained for its release to the environment.

Figure 3.2
Simplified Mechanical Flow Diagram of KPEG Field-Scale Treatment System at Guam, U.S.A.

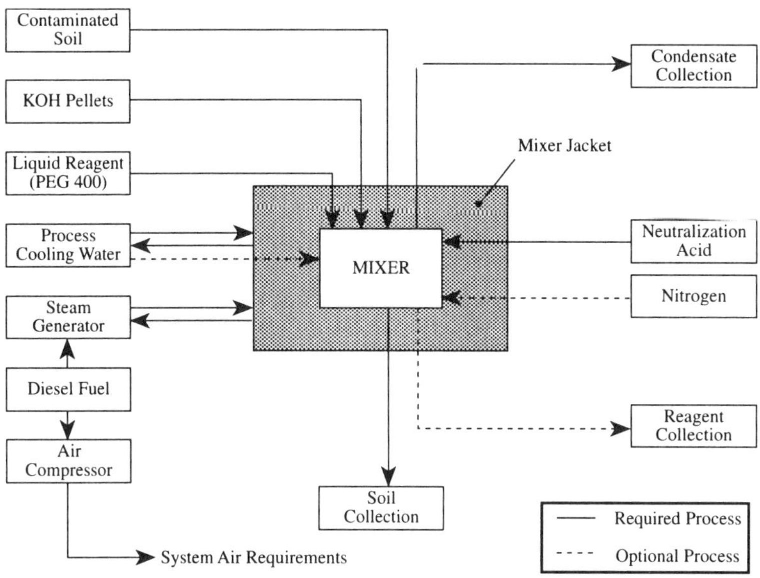

Reprinted by permission of the Hazardous Material Control Resources Institute from "Industrial Plant Expansion: Groundwater and Soil Cleanup" by J.E. Edwards and T. Bonham in the 5th National Conference on Hazardous Wastes and Hazardous Materials. Copyright 1988 by Hazardous Material Control Resources Institute.

The reactor was a 3,000-L (793-gal) jacketed vessel with a working capacity of 1,900 L (490 gal). It was equipped with a 75-hp motor and a gear box stirrer capable of providing mixer shaft speeds of 30 and 60 rev/min. A nitrogen purge was maintained on the system for safety. The reactor was heated by circulating heat transfer fluid through the jacket. All wettable parts of the system were made of 316 stainless steel. (Ferguson and Rogers 1990).

The results were similar to those of other processes; substitution of the aromatic chloride by the alkoxyethoxy moiety for the aromatic chloride occurs, except that, it is claimed, all chlorine atoms on the PCB were replaced. No proof of such complete substitution is given, although complete substitution is not necessary in order to convert the PCB into a unregulated form. The treated soil leaving the KPEG process was saturated with PEG after this demonstration. It was described by one observer as similar to quicksand.

3.1.1.2 GRC Process

The GRC Process (also called the alkaline polyethylene glycol (APEG) Plus Process) was developed by the Galson Remediation Company (Peterson 1986). It uses KPEG mixed with dimethyl sulfoxide (DMSO) as the reagent. The DMSO extracts the contaminant from the soil, reduces the PEG's viscosity, and catalyzes the substitution reaction (GRC 1992). The first system employing DMSO and KPEG was designed for liquids, PCB-contaminated transformer oil. A specially-designed second system was subsequently used to treat contaminated soils.

Before treatment, the site is prepared by excavating and stockpiling the contaminated soil. The contaminated soil is screened to less than 1.5 cm (0.6 in.) and sent through a shredder; larger particles are removed by the separator and washed. Screened particles are loaded and conveyed to a wet slurry mixing feed system where they are reacted with reagents (Peterson 1986).

The soil/reagent mixture is heated to 150°C (300°F) and reacted for several hours in the presence of a 45% aqueous solution of potassium hydroxide (1 part), polyethylene glycol (4 parts), DMSO (1 part), and the chloroaromatic contaminant. In clay soils, the recipe calls for 125 lb of reagent mixture per 100 lb of soil. If the contaminant soil is sandy, then 60 lb of reagent mix is used with 100 lb of soil. The level of contaminant, soil type, water content (which should be less than 30% in clay soil), and size of soil particles all affect the separating conditions.

After treatment is completed, free reagent is separated by decantation. This is followed, as shown in figure 3.3 (on page 3.13), by a series of washes with water to remove the residual KPEG reagent. Note that both PEG and KPEG are miscible with water. After washing, the clean soil is partially dried and discharged. The washwater is distilled to recover the PEG, DMSO, and potassium hydroxide. The residue from the distillation consists of the reaction products potassium chloride and hydroxylated(s) polychlorobiphenyls.

Figure 3.3
GRC Process Flow Diagram

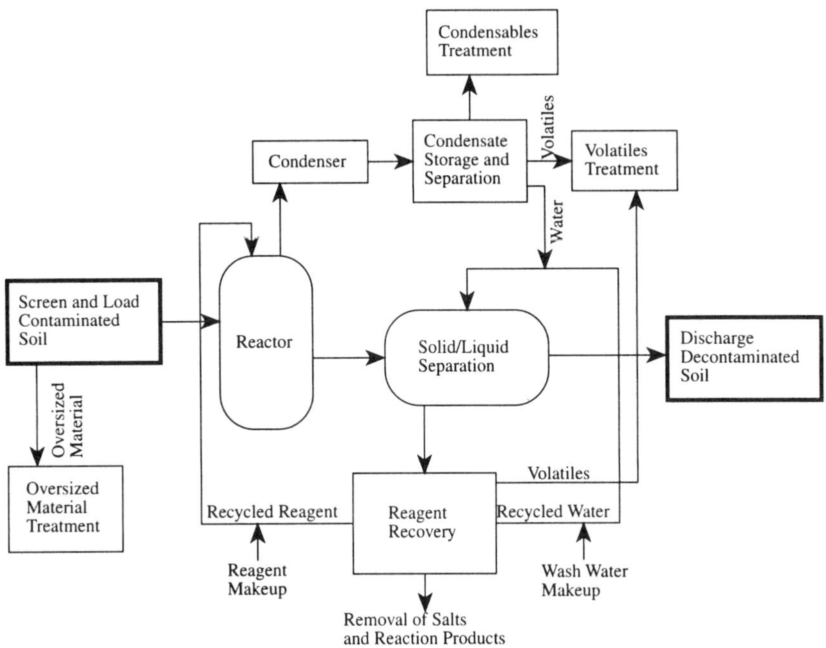

3.1.1.3 KGME/DECHLOR

The KGME/DECHLOR Process was developed by Chemical Waste Management (CWM) (Friedman and Halpern 1992a) to treat oils contaminated with PCBs, polychlorodibenzodioxins (PCDDs), and polychlorodibenzofurans (PCDFs). The process is coupled with CWM's X*TRAX Process to treat soil. X*TRAX is a thermal treatment system that removes oils and oily contaminants from soil and other solids. X*TRAX removes the contaminated oil from the soil and the oil is treated by the KGME/DECHLOR Process.

The KGME/DECHLOR Process employs the reaction of 2-methoxyethanol with either KOH in the presence or absence of an nonaqueous hydrogen donor (aprotic) solvent, forming potassium polyethylene glycol methylethers (GME) (KGME). The KGME reacts with haloaromatics, replacing one or more halogens with the 2-methoxyethoxy moiety and are converted to methoxyethoxy aryls. That is, the chlorine atoms attached to the aryl nucleus are replaced with the methoxyethoxy moiety, forming ethers at the aryl nucleus.

Details of the reaction are found in the CWM brochure (Friedman and Halpern 1992a, 1992b). To produce KGME, 2-methoxyethanol is reacted with potassium hydroxide in a suitable reaction vessel containing halogenated aryls at 110 to 150°C (230 to 300°F) during a one to four hour period. The treated oil phase, which contains the reaction products and traces of unreacted PCBs, is transferred to a storage tank. Samples are taken for analysis, and if the PCB concentration is unacceptable, the batch is returned to the reactor for a more vigorous treatment. If the PCB concentration is acceptable, the batch is discharged. The final disposition of the treated oil depends on whether other regulated Resource Conservation and Recovery Act (RCRA) hazardous constituents are present.

Aroclors, 1254 and 1250, are reduced from initial concentrations of 250,000 mg/L to <50 mg/L and sometimes to <5 mg/L. Lower Aroclors, e.g. 1016, 1242 and 1248, can also be transformed to non-PCBs, but generally require greater concentrations of reagent, higher temperatures, and/or longer reaction times.

Chemical Waste Management is currently using a pilot-scale KGME/DECHLOR system depicted in figure 3.4 (on page 3.15). The major components consist of a 380-L (100-gal) reactor, thermal fluid heater/chiller, a 1,900 L (500 gal) decantation tank, storage tanks of 1,900 L (500 gal) capacity, and pumps.

Chemical Waste Management indicates that the system converts PCBs into substituted ethers with methoxyethoxy moieties at somewhat greater efficiency than does KPEG at 115°C (239°F). But at a higher temperature, 150°C (300°F), there is only a very slight improvement. No data are available for KPEG reaction at 150°C. An important point is noted in the CWM report, which states:

> "The dehalogenation shown ..... was performed on a waste which contained 722 ppb of PCDDs and 2725 ppb of PCDFs prior to treatment. After 1.25 hours at 115°C (one equivalent KGME/equivalent Cl; 0.5 equivalent of 2-methoxy ethanol excess), the chlorinated dioxins and furans were reduced to

> below limits of detection (51 ppb). A similar treatment with KPEG resulted in a comparable destruction of dioxins; however, the levels of tetrachlorinated dibenzofurans were actually increased (from 30 to 251 ppb)."

It should be noted that methoxy ethanol is being tested as a possible teratogen. Larger molecules, such as methoxyethoxy-ethanol, are listed as irritants.

The KGME/DECHLOR Process is currently limited in application to liquid wastes with less than 25% water. Solid materials contaminated with PCBs require a pretreatment step, such as X*TRAX or soil washing, in order to remove the PCBs and to provide a treatable liquid matrix.

Figure 3.4
DECHLOR/KGME Process Flow Diagram

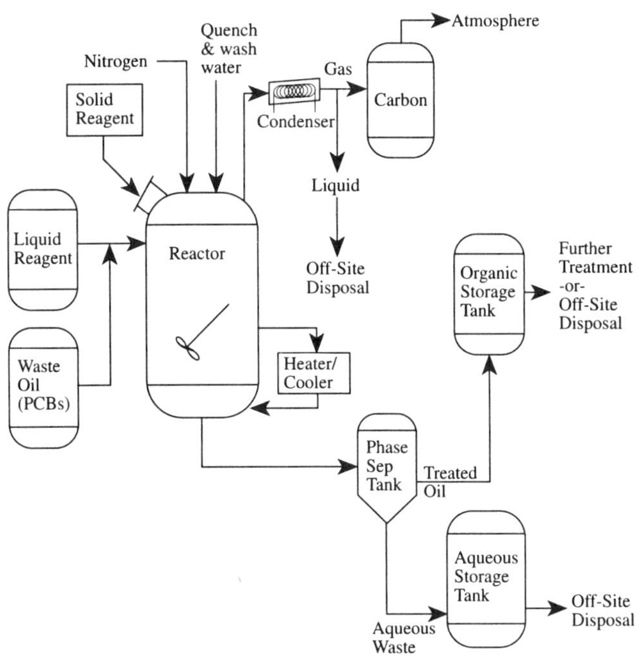

3.1.2 High-Temperature Substitution Processes

Two high-temperature substitution processes were evaluated — the ATP (SoilTech) Process, addressed here, and the BCD Process, an emerging technology, addressed in Appendix A.

The ATP Process, developed by SoilTech, Inc., was evaluated under the U.S. Environmental Protection Agency's (US EPA's) Superfund Innovative Technology Evaluation (SITE) program. It has treated 36,000 tonne (40,000 ton) of PCB-contaminated soil at the Wide Beach (Brant, NY) Superfund site and PCB-contaminated dredge spoils at Waukegan Harbor, Illinois as part of a demonstration. The process is based on a thermal desorption process patented by Taciuk (dePercin 1991; Taciuk 1979, 1981a, 1981b). The system can treat 9 tonne (10 ton) of soil per hour.

Figure 3.5 (on page 3.17) shows the Taciuk reactor, as depicted in the patent. Note that the referenced patent is for use of the reactor to extract oil from shale; the system used for remediation may differ in some details. Figure 3.6 (on page 3.18) is a simplified diagram of the reactor. See figure 3.7 (on page 3.19) for a flow diagram of the entire system.

The reactor is a rotating cylinder consisting of a core (called here "the reactor") surrounded by an annular space through which hot combustion gases flow. Crushed and sized contaminated soil, mixed with a solution of diesel fuel, recycle oil from the process, and APEG (whether sodium or potassium, is not specified) is fed to the reactor (on the left of figure 3.5 on page 3.17) (A). The solids flow to the far end (B), then drop into the outer chamber and flow back to the discharge. A fossil fuel burner (B) is mounted at the point where the solids drop into the annular chamber.

The reactor and the annular chamber are each divided into two thermal zones by a baffle. The region to the left of the baffle in the reactor, called the "preheat zone," is maintained at a temperature of 200 to 340°C (400 to 650°F). The region to the right of the baffle, called the "retort zone," is maintained at a temperature of 480 to 620°C (900 to 1,150°F). In the annular chamber, the region to the right of the baffle is called either the coking or combustion zone, depending on whether it refers to the portion of the kiln which is below or above the solids bed in the rotating annulus. The region to the left of the baffle is referred to as the cooling zone. The temperature in the combustion zone (and presumably coking zone) of the reactor is 650 to 760°C (1,200 to 1,400°F), and in the cooling zone, 260 to 430°C (500 to 800°F).

Figure 3.5
Rotating Kiln Processor

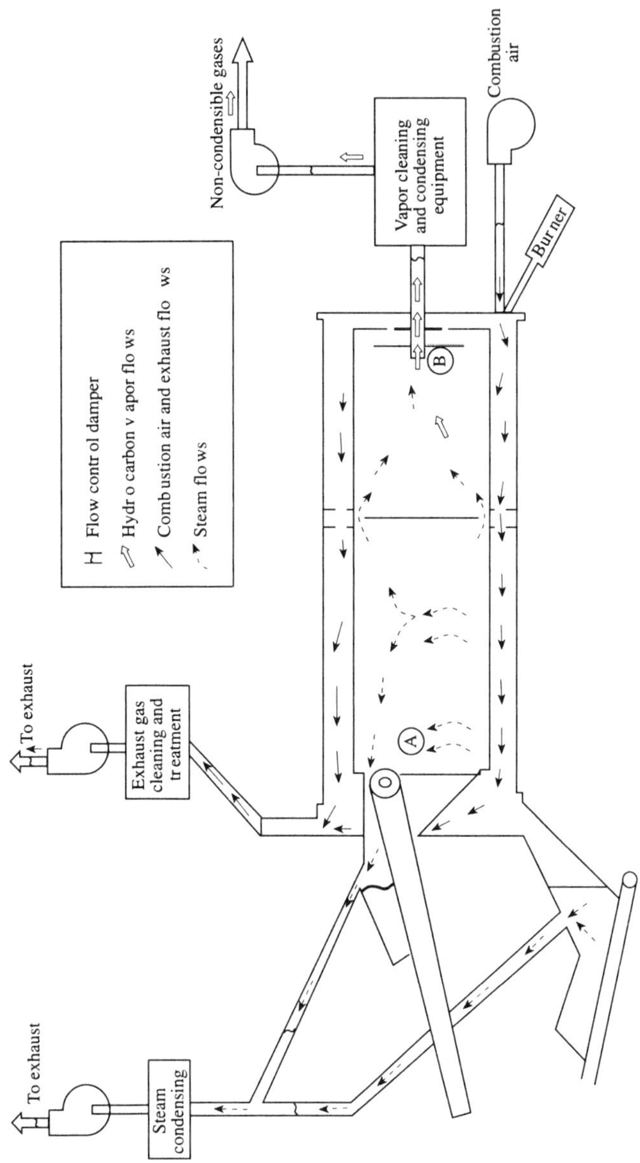

Figure 3.6
Simplified Sectional Diagram Showing the Four Internal Zones — ATP

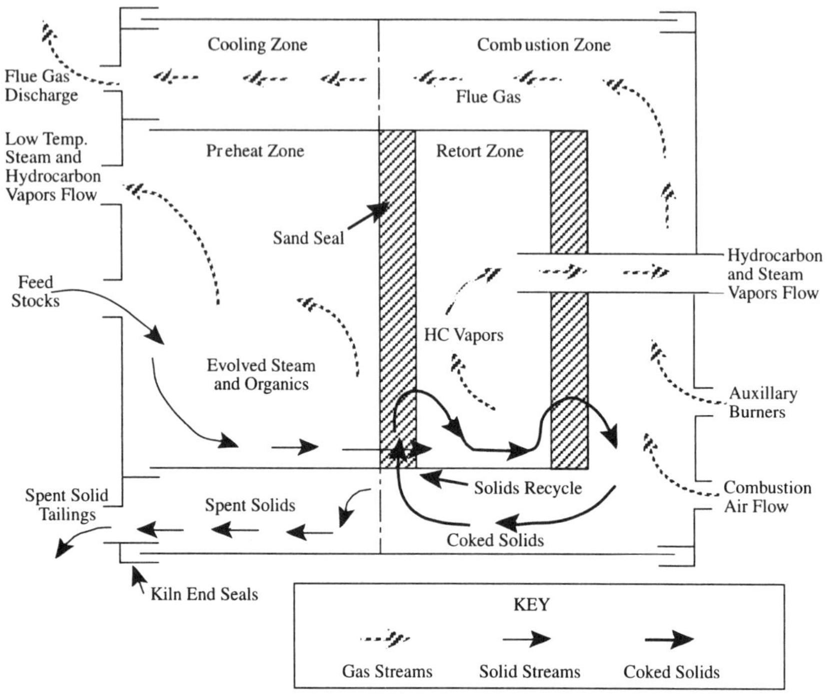

The flow of solids and gases in the reactor is complex. The organic constituents vaporize in the preheat zone and those that do not vaporize crack to lower molecular weight compounds in the retort zone. In addition, the reagent, by dechlorinating the PCB, also contributes to its destruction. Offgases from the retort zone flow out of the reactor to condensers and air pollution control equipment, as shown in figure 3.6. The condensate is physically separated into an aqueous phase and heavy and light nonaqueous phases. The nonaqueous phase is spread on the soil feed and recycled back into the reactor. The aqueous phase is treated by a carbon adsorption system. Spent carbon from the two systems is recycled back into the reactor.

The hot, granular solids exiting the reactor pass through the fossil fuel burner flame before dropping into the annular chamber. The burner ignites residual organics on the soil. Gases from the burner mix with those of the

Figure 3.7
Process Flow Diagram, ATP SoilTech

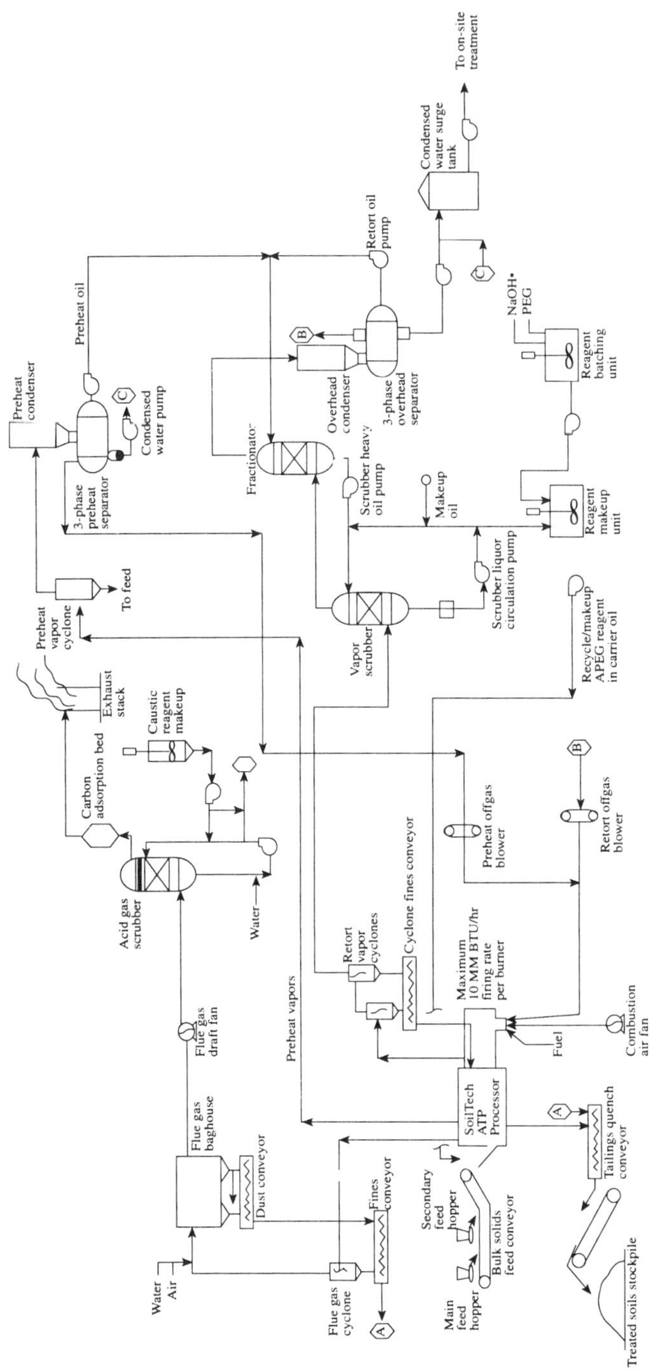

Source: US EPA 1992c

burning organics in the annular region and pass into the combustion zone of the furnace.

Figure 3.7 (on page 3.19) is a process flow diagram of the complete system. The ATP Process combines features of thermal desorption,[1] thermal destruction (incineration), and chemical reaction in one unit.

Forty thousand tons of PCB-contaminated soil was treated at the Wide Beach Development Superfund site. The concentration of PCBs was reduced from the 100 to 600 mg/L range to less than 10 mg/L. The treatment was performed in 1990-91 at a cost of $245/ton. The only question that might be asked about the process is what fraction of destruction of PCBs took place because of thermal decomposition versus chemical decomposition? The process achieved an acceptable level of destruction of the target contaminants. The physical condition of the treated soil was not reported. However, because the treatment process is basically a thermal system, the product is, most likely, a dry, granular solid.

3.2 *Oxidation Process Descriptions*

Chemical oxidation processes are used in many industrial processes (RadTech 1992) and in the treatment of potable water to remove a broad range of natural and synthetic organic compounds. Many of these technologies are commercially available and can be readily applied in the remediation of groundwater and even soil. Oxidation processes fall into three broad categories: (1) chemical oxidation, (2) photodegradation/photolysis, and (3) a combination of chemical oxidation and photolysis. The emerging technologies, discussed in Appendix A, do not clearly fit these categories.

Oxidation processes have been used exclusively to degrade organic compounds in aqueous media and some soils; however, it would appear possible to use the technique also to change the oxidation state of metals and convert them to less toxic or less soluble forms that will precipitate.

1 See the monograph in this series, *Innovative Site Remediation: Thermal Desorption*, wherein the ATP Process is addressed as a thermal desorption process.

3.2.1 Chemical Oxidation Processes

Chemical oxidation processes use merely a chemical oxidizing agent to react with the contaminant. The following are commonly used as oxidizing agents:

- hypochlorite, either sodium or calcium ($NaOCl$ or $Ca(OCl)_2$);
- hydrogen peroxide; and
- ozone.

Hypochlorite treatment is a well-established technology that has been used extensively to destroy pathogens in drinking water. A case history of its use in remediation to destroy cyanides is given in Appendix B. Its use, however, will likely remain limited because it is relatively expensive and because it can convert organic constituents to traces of chloromethanes and chlorethanes.

Ozone and hydrogen peroxide are excellent oxidizing agents that can be used somewhat interchangeably for the destruction of organic compounds. They are both strong oxidizing agents capable of destroying most nonhalogenated and some halogenated compounds in aqueous media. For example, both have been commonly used to destroy cyanides in wastewater. Similarly, they can be used to destroy low concentrations of easily oxidizable organic compounds in groundwater.

As compounds involved become refractory (hard to destroy or chemically react) or as the concentration increases, the amount of oxidizing agent required becomes high. Although ozone and hydrogen peroxide are both oxidizing agents, they are not completely interchangeable. The chemical pathways they follow in oxidizing organic compounds are somewhat different. This is especially true in the presence of UV light, as explained in the next subsection. An understanding of the fundamental chemistries involved is necessary to design an efficient treatability study, and in most cases the treatability studies should include consideration of ozone and peroxide both individually and in combination.

3.2.2 UV Photodegradation/Photolysis

Ultraviolet light can initiate oxidation reactions either through simple photodegradation, with low-intensity UV light, or through photolysis, with high-intensity UV light. A major potential application of photolysis is the use of natural sunlight to degrade organics in soil and waters. Photolysis is a true oxidative process. The high-intensity UV light forms OH^- radicals that attack and oxidize the organic contaminants.

While photodegradation can be used to destroy some compounds, total destruction of most compounds has not proven feasible. Dissociation of chemical bonds occurs only when the UV light wavelength matches the absorption bands of the contaminants. Each contaminant has a unique optimal wavelength for photodissociation; and as a contaminant degrades partially, the wavelength that will degrade the product molecule is usually different from that which degrades the original compound.

Another difficulty with solar photodegradation is that the absorption bands and wavelengths are usually not the same for the various by-products of photooxidation and therefore, the reaction products are themselves refractory. A good example of this difficulty is the "pinkwater" that results from exposure of trinitrotoluene (TNT) production wastewater to sunlight. The pinkwater contains partial products of oxidation of the original contaminants and is in its own right an environmental problem.

The type of UV light source used for photodegradation is critical to the process's success. This is equally true for simple photodegradation processes or advanced oxidation processes, discussed in Subsection 3.2.3.

Photodissociation of organic compounds by direct photolysis requires photon energies from 4 eV to 7 eV (1 eV = 1.6×10^{-19} Joules), which corresponds to wavelengths from 300 nm to 175 nm. Complex structures and mixtures of organic compounds require a very dense structure of UV emission lines to be completely degraded to innocuous products because each molecule has a set of unique absorption bands that can vary in width (1 to 20 nm) and dictate an optimal wavelength for photodissociation. For example, benzene's molar extinction coefficient is 47,000 at a wavelength of 184 nm, 7,000 at 202 nm, and at wavelengths between 230 to 270 nm, benezene absorbs light relatively weakly with an extinction coefficient, (ε = 300) (see Subsection 3.4.2.1). In contrast, acetone has strong absorbance at a wavelength 220 nm (ε = 16,000), but is very weak at 318 nm (ε = 30) (Wekhof, 1991).

First generation UV light sources are able to generate only a few lines in the far UV region. Mercury vapor lamps emit their strongest emission at 254 nm. Medium- to high-pressure mercury vapor lamps with input powers greater than 80 watt/cm have additional strong lines at 248, 265, 280, 297, and 302 nm. Given this limited availability of emission lines, it is apparent that even high-power, medium pressure mercury vapor lamps are incapable of cleaving bonds in contaminants such as benzene.

Metal halide lamps, which are medium- or high-pressure mercury lamps containing dopants, such as iodides and magnesium, have a higher emission line density in the far UV region. Dopants, which can be selected to fill specific wavelength gaps between the mercury lines, allow customizing the spectrum of the lamp to a particular application. These types of lamps are far more efficient than first generation mercury vapor lamps, which require the addition of an oxidant for efficient toxic degradation. The efficiency of doped mercury lamps, however, can be enhanced through the addition of peroxide and/or a photocatalyst.

Even mercury lamps containing dopants lack a sufficient number of UV lines to insure complete destruction of mixtures of toxic chemicals. If doped mercury lamps are considered second generation technologies, continued innovation has brought about the development of a third generation lamp that provides an UV continuum with pulsed devices. The most simple and cost-effective types of broad UV spectrum lamps for environmental applications are xenon and custom-built flashlamps (Wekhof, 1991). Operation of the xenon flashlamp entails converting the xenon gas to a plasma by a short pulse of electric current. In order to increase the UV output below 300 nm, xenon flashlamps must be operated outside their rated range of current density (6 to 14 kA/cm^2 vs. 1 to 5 kA/cm^2), which will significantly shorten lamp life by a factor of 100 or more.

In water, direct photolysis occurs readily of compounds that have sufficient solubilities and are exposed to specifically energetic light. The effects of pH and ionic strength on the process are relatively minor. Organic destruction in air, however, is about 5 times faster than in water (Wekhof, 1991). This is due to the following factors:

- Water absorbs UV light more than does air;
- Recombination of photoproducts to yield the parent compound is less in air than in water because of diffusional properties; and
- Ozone is formed in the gaseous system because of the presence of oxygen.

Although UV radiation does not penetrate solid matrices, such as soils, direct or indirect photolysis can be used effectively in soil systems by creating an organic or surfactant film on the surface of particles to solubilize the contaminants at the interface. In addition to transporting contaminants to the surface, organic solvents/surfactants may also play the role of H^+ -donors in the degradation pathway. In addition, in soil irradiation shallow depth and periodic mixing are necessary.

3.2.3 Advanced Oxidation Processes

Advanced oxidation processes, based on free radical, chain reaction chemistry, combine ozonation with UV photolysis and hydrogen peroxide. Some compounds (e.g., methanol) that are resistant to destruction or are only slowly destroyed by UV radiation alone or with the addition of either O_3 or H_2O_2 are rapidly destroyed to a far greater extent in the presence of all three agents, suggesting that a synergy exists in the chemistry of this process.

Over the past 15 years, the oxidation of a large number of organic contaminants in wastewater throughout the use of a combination of ultraviolet light and hydrogen peroxide has been examined. While ultraviolet light and hydrogen peroxide are each capable of causing some degradation of organic materials by themselves, the synergistic effect of combining the two methods expands their utility and effectiveness (Ansari, Kahn, and Ali 1985). The complete mineralization of 2,4-Dinitrotoluene (2,4-DNT) proceeds through sidechain oxidation, hydroxylation of the benzene ring, and benzene ring cleavage in the presence of ultraviolet light and hydrogen peroxide. The degradation of 2,4-DNT proceeds rapidly at H_2O_2:2,4-DNT mole ratios above 13:1 (Ho 1986). Initial concentrations of 75 mg/L of 2,4-DNT in water can be reduced to below detection limits of 1 mg/L in 45 minutes or less. During the oxidative degradation of 2,4-DNT, the pH of the reaction medium is reduced from approximately 6.4 to approximately 2.6.

Guittonneau et al. (1990) showed that p-chloronitrobenzene undergoes oxidative degradation with UV/H_2O_2 treatment at ambient temperatures. Sundstrom and Klei (1986) demonstrated the oxidative degradation, under the influence of UV/H_2O_2 treatment, of a wide range of organic compounds in water, including o-dichlorobenzene, m-dichlorobenzene, chlorobenzene, phenol, toluene, carbon tetrachloride, chloroform, ethylene dibromide, and dichloromethane. In addition, the reduction of the pentachlorophenol content of contaminated groundwater from 10 mg/L to less than 1 mg/L has also been described (Edwards and Bonham 1988). The partial oxidation of arginine to a variety of amino acids and urea was reported by Ansari, Kahn, and Ali (1985). Although 1,1,1-trichlorethane and Freon 113 do undergo oxidative degradation when treated with the UV/H_2O_2 system, overall degradation of these compounds is significantly slower than the degradation of other organics such as perchloroethylene (Camp 1991).

Since the chemistry of advanced oxidation processes is essentially that of the hydroxyl free radical, they are extremely effective treatment processes for most contaminants. For example, the dechlorination of a highly chlorinated aliphatic compound, chlorendic acid, can be accomplished with ozone, but the rate is enhanced significantly under conditions which favor the formation of hydroxyl radicals: UV light, high pH, and low bicarbonate concentration (Stowell and Jensen 1991). The degradation rates of volatile organochlorine compounds (trichloroethane, trichloroethylene, and tetrachloroethylene) by ozonation were increased by a factor of 10 with UV irradiation (Kusakabe et al. 1991).

Recently, two advanced oxidation processes that appear to be able to treat a range of refractory compounds, including pentachlorophenol, were introduced. They are the Rayox (Solarchem Environmental Systems, San Diego, Calif.) and Ultrox (Santa Ana, Calif.) Processes. The fundamental feature of both is the use of a combined UV/ozone or UV/ozone/H_2O_2 to generate OH• radicals.

Rayox Process. The Rayox Process uses a proprietary reactor design that improves mass and radiation transfer, permits staged treatments, and enables automatic cleaning of the UV components. The process uses proprietary UV light sources termed by the vendor as Solarchem UV lamps. These lamps have a range of intensities and emission characteristics matched to the pollutants treated. The process also utilizes proprietary catalysts termed the ENOX catalysts, to enhance the production efficiency of OH• radicals and oxidation selectivity. The Rayox Process has twice the number of controllable process variables as early oxidative process technology, and the user can select process conditions to favor radical reactions, direct photolysis reactions, or a combination of the two. The process train can use either a batch or continuous flow reactor depending on type and amount of contamination, flow rate, and degree of removal desired.

According to the vendor, the Rayox UV/ozone/H_2O_2 Process has been shown to be effective in treating most halogenated and volatile compounds, PCBs and dioxins, pesticides, organic corrosive chemicals, nitrogenated aromatic and aliphatic compounds, cyanides, and iron. The process has been applied in treating the following:

- gasoline in groundwater for destruction of benzene, toluene, ethylbenzene, xylene (BTEX);
- multi-contaminant leachates (carbon tetrachloride, benzene, toluene, xylene (BTX), chlorinated solvents, ketones, chloroform, etc.);
- wood preservation industrial wastewaters (phenols, PCP), dioxins, polyaromatic hydrocarbons (PAHs));
- chlorinated and nonchlorinated solvents (trichloroethylene (TCE), tetrachloroethene (PCE), dichloroethane (DCA), 1,4-Dioxane, etc.); and
- explosives in process waters (TNT, DNT, nitroglycerine (NG), ethylene glycol dinitrate (EGDN), etc.).

When used at an US EPA superfund site, >99% volatile organic compound (VOC) destruction was achieved, and in a groundwater remediation, >99.99% BTX degradation was effected (US EPA 1990). Successful treatment of process water and groundwater containing mixed pesticides, PAHs from coal gasification, complexed (including iron) cyanides, dyes, and bleach plant effluent has been reported by the manufacturer.

No schematic or operational information was received for the Rayox Process, but it appears to be similar in its major features to the Ultrox Process, described below.

Ultrox Process. The Ultrox Process uses a combination of ozone, hydrogen peroxide, and UV irradiation to treat a variety of organic constituents in aqueous streams. See figure 3.8 (on page 3.27). The system can be obtained in a variety of sizes with reactors ranging in capacity from 1,100 to 14,800 L (300 to 3,900 gal). The reactor is made of stainless steel. It contains vertical, low temperature mercury lamps inside quartz tubes. Four to eight reactors can be staged in series, the number depending on treatment conditions. Lamps may be mounted in all or only some of the reactors. The process can operate under intermittent, continuous, or batch flow conditions and may be fully automated, with periodic monitoring.

The reactor uses ozone and/or hydrogen peroxide. The ozone is generated either from air (producing a stream containing 2% ozone by weight) or

Figure 3.8
The Ultrox System

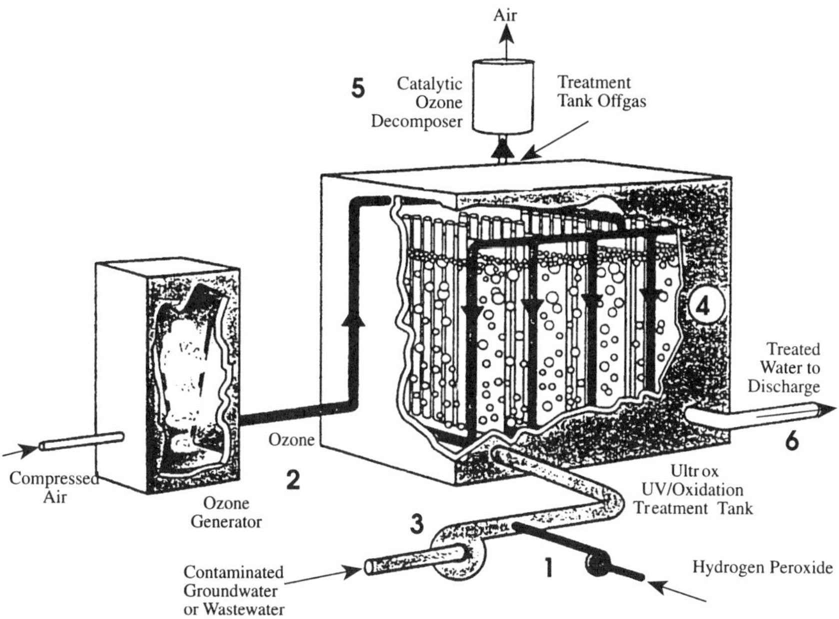

1 Hydrogen peroxide is combined with contaminated water.
2 Ozone is generated and injected into the treatment tank.
3 Contaminated water is pumped to the treatment tank and irradiated with ultraviolet light. The light reacts with the ozone gas and hydrogen peroxide, producing hydroxyl radicals which destroy organic contaminants.
4 Water flows from left to right through a series of treatment chambers.
5 Residual ozone in the offgas is converted to oxygen by a catalytic decomposer, eliminating any release of ozone.
6 Treated water flows to discharge.

oxygen (yielding 6% ozone by weight) in the range of 4 to 44 kg/day (10 to 100 lb/day). The ozone is bubbled into the base of the reactor. Hydrogen peroxide may replace ozone, or be used in combination with ozone and is directly metered into the influent line. The offgas from the reactor is collected for destruction of residual O_3 and VOCs by a patented catalytic process, D-TOX™. The D-TOX™ uses O_3 to oxidize the susceptible organics and also uses two additional catalysts to destroy remaining VOCs and residual O_3. A sorber made of a mixture of bases is located at the end of the catalytic train to remove any remaining organic material or acids.

The Ultrox Process was demonstrated in 1988 under the SITE program (US EPA 1989). The treated wastewater contained 44 organic contaminants, three of which, TCE, 1,1-DCA, and 1,1,1-trichloroethane (1,1,1-TCA), were chosen as indicator parameters for the test. Efficiency in removing the TCE was about 99%, the 1,1-DCA, about 58%, and the 1,1,1-TCA, about 85%. Measured efficiencies in removing the total VOCs were about 90%.

Both chemical oxidation and stripping resulted in removal of some compounds from the water phase. Stripping accounted for 12 to 75% of the total removal of 1,1,1-TCA and for 5 to 44% of 1,1-DCA. Stripping accounted for less than 10% of the TCE and vinyl chloride removed and negligible amounts of other VOCs present.

The ozone destruction system reduced ozone concentrations in the gas stream to less than 0.1 ppm and reduced the VOCs, which were stripped into the gas stream, to nondetectable concentrations. Very little removal of total organic carbon (TOC) was observed, indicating partial oxidation of organics without complete conversion to carbon dioxide and water.

According to the vendor, the Ultrox AOP is capable of treating contamination typical of that experienced in the petroleum industry (leaking fuel storage tanks, refinery equipment leaks, spills at transfer terminals, and pipeline failures) and is effective in treating compounds, such as benzene, toluene, xylene, ethyl benzene, and methyl-t-butyl ether (MTBE). As to the chemical industry, the Ultrox system is capable of degrading benzene, phenols, TCE, PCE, and chlorinated or phosphated pesticides. This technology is also applicable in treating compounds typically found in contamination experienced in the aerospace industry, such as Freon 113. Contamination commonly experienced in the woodtreating industry, such as creosote, PAHs, and dioxins can be handled by the Ultrox system. In groundwaters, the following contaminants can be degraded by the Ultrox system:

- benzene
- toluene
- creosote
- 1,2-dichloroethylene
- perchloroethylene (PCE)
- xylene
- pentachlorophenol (PCP)
- bis(2-chloroethyl)ether

- dichloroethylene
- dioxins
- dioxanes
- Freon 113
- methylene chloride
- methyl isobutyl ketone
- polychlorinated biphenyls (PCBs)
- tetrachloroethylene
- pesticides
- polynuclear aromatics (PNA)
- 1,1,1-trichloroethane (1,1,1-TCA)
- trichloroethylene (TCE)
- tetrahydrofuran
- vinyl chloride
- methyl-butyl ether (MTBE)

Industrial wastewaters containing the following compounds are also effectively treated:

- amines
- analines
- chlorinated solvents
- chlorobenzenes
- creosotes
- complex cyanides
- hydrazine compounds
- isopropanol
- methyl ethyl ketone
- methyl isobutyl ketone
- methylene chloride
- pesticides
- phenol
- Royal Dutch Explosive, cyclo-1,3,5-trimethylene-2,4,6-trinitramine, or cyclonite (RDX)
- trinitrotoluene (TNT)
- polynitrophenols

3.3 Precipitation Processes

Chemical precipitation involves transforming a soluble compound into an insoluble form through the addition of chemicals, such that a supersaturated environment exists (i.e., the solubility product is exceeded). It should be noted that there is a fine distinction between chemical precipitation and stabilization/solidification (S/S) operations. In S/S operations, the contaminants are incorporated into a cement-like matrix, rendering the contaminants less prone to leaching. Sludges are chemically treated by mixing a binder material to improve the physical and chemical stability of the sludge. Materials such as portland cement, silicates, pozzolanic materials, and fly ash have been used as S/S agents. The purpose of S/S technologies is to minimize the leaching potential of the contaminants.[2] Similarly, this is the goal of chemical precipitation operations, that is, to make the contaminant less soluble. S/S techniques are used to immobilize heavy metals, but they have also been used to immobilize organic contaminants as well. In general, organics with low water solubility are immobilized fairly well through S/S operations, while higher solubility organics are not (Connor 1990). Chemical precipitation techniques are rarely used to precipitate organic compounds from solution, although organics can adsorb onto precipitate forms, such as hydrous metal oxides.

Chemical precipitation is the most common technique used for treatment of metal-containing waters (US EPA 1980; Peters, Ku, and Bhattacharyya 1985; Patterson 1988; Patterson and Mincar 1975). Oxidation/reduction plus precipitation being a closely-related technique is also used (Patterson 1988). Alternative techniques, including selective sorption/desorption and differential precipitation, have focused primarily on opportunities for recovery of metals and sludge beneficiation and extraction (Patterson 1988). Clifford, Subramonian, and Sorg (1986) cite the following advantages of precipitation/coprecipitation contaminant removal processes:

- low cost for high volume;
- often improved by high ionic strength; and
- reliable and well-suited for osmotic control.

Limitations include the following:

- stoichiometric chemical addition requirements;
- high water content sludge must be disposed;
- part per billion effluent contaminant levels may require two-stage precipitation;
- processing is not readily applied to small, intermittent flows; and
- coprecipitation efficiency depends on initial contaminant concentration and surface area of the primary floc.

---

2 See the monograph in this series, *Innovative Site Remediation: Stabilization/Solidification.*

Precipitation can be broadly divided into two categories: chemical precipitation, and coprecipitation/adsorption. Chemical precipitation is a complex phenomenon resulting from the induction of supersaturation conditions. Precipitation proceeds through three stages: nucleation, crystal growth, and flocculation. Metal salt solubility can be predicted (at equilibrium) from thermodynamic calculations. These thermodynamic calculations cannot assess the kinetic rate, the influence of precipitate induction parameters, or the degree of supersaturation required to induce nucleation. It is important to note that the stability constants reported in the technical literature can vary by several orders of magnitude. See, for example, table 3.4 (on page 3.32). Patterson (1988) points out that the shape of the cadmium hydroxide solubility curve (as a function of pH) can vary significantly, depending on the particular stability constants chosen.

Chemical precipitation processes offer significant potential for removing soluble ionic species from solution, particularly heavy metals. The technique is not generally applicable in treatment of contaminated soils. It can, however, be used to treat industrial wastewaters and contaminated groundwaters (ex situ). This technique is applicable for removing soluble ionic species from aqueous solutions. Chemical precipitation can also be used as a pretreatment technique to remove heavy metals from solution before biodegradation of hazardous organic compounds.

There are five basic precipitation techniques that can be used to remove heavy metals from solution. Each are discussed in the following five subsections.

3.3.1 Hydroxide Precipitation

In hydroxide precipitation, heavy metals are removed by adding an alkali, such as caustic or lime, adjusting the wastewater pH to the point where the metal(s) exhibits minimum solubility. In general, the solubilities of metal hydroxides in solution decrease with increasing pH to a minimum value beyond which (the isoelectric point) the metals become more soluble because of their amphoteric nature. See figure 3.9 (on page 3.33). Newkirk, Warner, and Barros (1981) observe that the minimum solubilities as quantified under ideal conditions differ considerably with those observed in actual practice because of the influences of complexing agents (and other contaminants that may be present), temperature, and ionic strength. Further, Bowers, Chin, and Huang (1981) observe that in heterogeneous sys-

Table 3.4
Logarithm of Stability Constants Reported for Cadmium

Stability Constant	Smith and Martell (1976)	Baes and Mesmer (1976)	Sillen and Martell (1971)	Snoeyink and Jenkins (1980)	Dean (1979)	Sawyer and McCarty (1978)
K_1	3.90	3.92	5.00	4.16	4.17	6.08
K_2	7.70	7.65	8.90	8.40	8.33	8.70
K_3	8.75	8.70	11.60	9.10	9.02	8.38
K_4	8.70	8.65	---	8.80	8.62	8.42
K_{sp}	-14.65	-13.65	-13.60	-13.60	-13.60	(-13.60)

Reference Source: columns 2–7.

Chemical Reactions Involved:

$Cd(OH)_{2(s)} \Longleftrightarrow Cd^{+2} + 2\,OH^-$

$Cd^{+2} + OH^- \Longleftrightarrow CdOH^+$

$Cd^{+2} + 2OH^- \Longleftrightarrow Cd(OH)_2^0$

$Cd^{+2} + 3OH^- \Longleftrightarrow Cd(OH)_3^-$

$Cd^{+2} + 4OH^- \Longleftrightarrow Cd(OH)_4^{-2}$

Solubility Constant Expression:

$K_{sp} = [Cd^{+2}][OH^-]^2$

Stability Constant Expressions:

$$K_1 = \frac{[CdOH^+]}{[Cd^{+2}][OH^-]}$$

$$K_2 = \frac{[Cd(OH)_2O]}{[Cd(OH)^o][OH^-]} = \frac{[Cd(OH)_2O]}{K_1[Cd^{+2}][OH^-]^2}$$

$$K_3 = \frac{[Cd(OH)_3^-]}{[Cd(OH)_2O][OH^-]} = \frac{[Cd(OH)_3^-]}{K_1K_2[Cd^{+2}][OH^-]^3}$$

$$K_4 = \frac{[Cd(OH)_4^=]}{[Cd(OH)_{3^-}][OH^-]} = \frac{[Cd(OH)_4^=]}{K_1K_2K_3[Cd^{+2}][OH^-]^4}$$

Adapted from Patterson 1988

tems, coprecipitation and complexation of more than one species may be occurring.

The metals precipitate as metal hydroxides and can be removed by flocculation and sedimentation/filtration operations. The extent of precipitation depends on the solubility product (K_{sp}) of the metal hydroxide and the equilibrium (stability) constants, K_i's, of the metal hydroxyl constants, plus the stability constants for other complexing agents that may be present (ethylenediamine tetraacetic acid (EDTA), nitrilotriacetic acid (NTA), citrate, tartrate, gluconic acid, cyanide, ammonia, etc.). The effectiveness of the solid/liquid separation is heavily dependent on the physical properties

Figure 3.9

Solubilities of Metal Hydroxides and Metal Sulfides as a Function of Solution pH

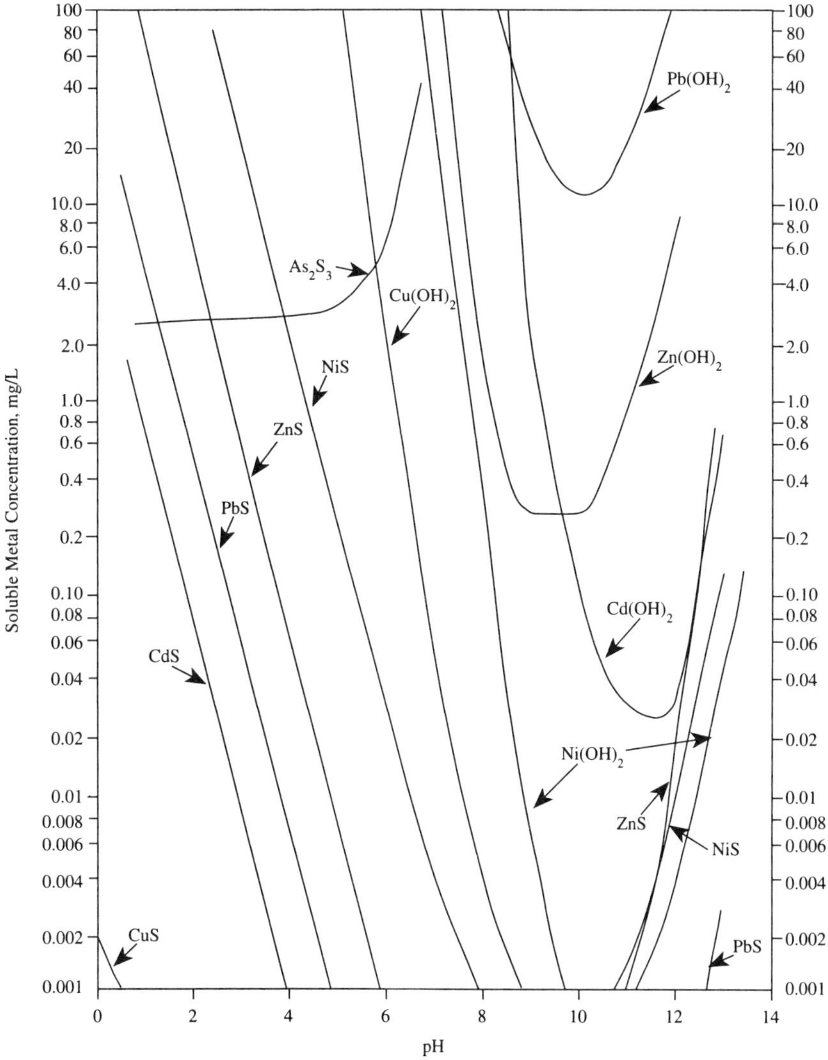

(size, density, etc.) of the metal hydroxide precipitates. Widespread use of this technique is due to its relative simplicity, low cost of precipitant (lime), and ease of pH control (Peters, Ku, and Bhattacharyya 1985; Peters and Ku 1987). See table 3.5 for a summary of advantages and limitations of metal hydroxide precipitation. Clifford, Subramonian, and Sorg (1986) observe that a staged precipitation process can be used for mixed-metal wastes because of the variation in the pH of their minimum hydroxide solubilities.

Arumugam (1976) studied hydroxide precipitation for recovery of chromium from spent tanning liquor. The process was found to be the cheapest for the removal and recovery of chromium. The optimum pH for maximum removal with lime was pH 6.6; removal of chromium exceeded 98% at that pH. The precipitated chromium hydroxide was separated by settling, filtra-

Table 3.5
Advantages and Limitations of Metal Hydroxide and Metal Sulfide Precipitation.

Advantages of Hydroxide Precipitation

- Ease of automatic pH control
- Well proven and accepted in industry
- Relatively simple operation
- Low cost precipitant (lime)

Limitations of Hydroxide Precipitation

- Hydroxide precipitates tend to resolubilize if the solution pH is changed
- The removal of metals by hydroxide precipitation of mixed metal wastes may not be effective because the minimum solubilities for different metals occur at different pH conditions
- The presence of complexing agents has an adverse effect on metal removal
- Chromium (VI) is not removed by this technique
- Cyanide interferes with heavy metal removal by hydroxide precipitation
- Hydroxide sludge quantities can be substantial and are generally difficult to dewater due to their amorphous particle structure
- Little metal hydroxide precipitation occurs at $pH<6$
- Processing is not stable for large flow and concentration variations in the influent
- Start-up and shutdown times are longer than those for packed-bed and membrane processes

Advantages of Sulfide Precipitation

- Attainment of a high degree of metal removal even at low pH ($pH \approx 2$ to 3)
- Low detention time requirements in the reactor due to the high reaction rates of sulfides
- Feasibility of selective metal removal and recovery exits
- Metal sulfide sludge exhibits better thickening and dewatering characteristics than the corresponding metal hydroxide sludge
- Metal sulfide precipitation is less influenced by the presence of complexes and chelating agents than is the corresponding metal hydroxide precipitation
- Metal sulfide sludge is reportedly three times less subject to leaching at pH 5 as compared with metal hydroxide sludge (Whang et al. 1981.)
- Metal sulfide sludges generally have smaller volumes and are easier to dewater than the corresponding metal hydroxide sludge

Limitations of Sulfide Precipitation

- Potential for H_2S gas evolution
- Possibility of sulfide toxicity
- Process is relatively complex and expensive as compared with hydroxide precipitation

tion, and then was redissolved in sulfuric acid to form chromium sulfate, which can be recycled for further tanning. Lime was more economical than other alkalies (NaOH, Na_2CO_3, and NH_4OH).

Sheffield (1981) investigated lime precipitation for removal of copper, iron, nickel, chromium, and lead. These metals, from electroplating shops, were successfully removed by precipitation with a hydroxide (such as lime) or soda ash with addition of sulfate or sulfide to enhance removal of the copper/iron complexes.

The batch precipitation of cadmium, zinc, and nickel both by hydroxide and sulfide precipitation for various pH conditions, reaction times, and type and concentration of complexing agents were studied. The complexing agents investigated were ammonia, citrate, phosphate, tartrate, and EDTA. The metal hydroxide precipitates tended to be amorphous and colloidal, causing the resulting sludge to be voluminous. The presence of complexing agents severely inhibited metal hydroxide precipitation (Peters and Ku 1985). Generally, higher pH conditions enhanced the nucleation rate and improved the resulting particle size distribution (Peters and Ku 1985; Peters et al. 1984; Peters, Ku, and Bhattacharyya 1984; Ku 1986). In the absence of chelating agents, extremely low residual zinc and cadmium concentrations (Zn <0.5 mg/L, Cd <0.3 mg/L) were obtained.

Hydroxide precipitation of heavy metals is well suited for automatic pH control and is an effective treatment technique in industry. As an example of the effectiveness of hydroxide precipitation, removal efficiencies exceeded 98% for Cd^{2+}, Pb^{2+}, and Cr^{3+} when spiked well waters and river waters were used (US EPA 1978; Sorg, Csanady, and Logsdon 1978; Sorg 1979).

Rabosky and Altares (1983) presented a case history of full-scale wastewater treatment for a small chrome-plating shop. Caustic soda was used to adjust the wastewater pH to between 9.5 and 10.0 in order to precipitate the metals as metal hydroxides. For the copper-containing wastewaters, the precipitation was performed at pH 10.5 using NaOH. See table 3.6 (on page 3.36) for typical results.

3.3.2 Carbonate Precipitation

Carbonate precipitation of heavy metals has been shown to be an effective treatment alternative to hydroxide precipitation. Carbonate precipita-

Table 3.6
Results After Treatment of Plant Wastewaters

Parameter	Concentration, (mg/L)	
	Before Treatment	After Treatment
Nickel	18.9	0.63
Copper	5.7	0.89
Lead	0.20	0.02
Zinc	1.35	0.06
Iron	8.4	2.8
Manganese	1.45	0.05
Chromium (total)	8.5	0.04
Chromium(+6)	4.5	<0.01
Cyanide	11.1	<0.001
Oil and Grease	1-10	1.2
Phenol	<0.001	<0.001
Suspended Solids	6-10	8.0

Adapted from Rabosky and Altares 1983

tion can be accomplished using soda ash (sodium carbonate). Carbonate precipitation has the following advantages over conventional hydroxide precipitation (Patterson, Allen, and Scala 1977; Clifford, Subramonian, and Sorg 1986):

- Optimum treatment occurs at lower pH conditions;
- Metal precipitates are reported to be denser than the liquid, facilitating solids separation; and
- Sludges have better dewatering characteristics.

Sodium bicarbonate can be used also to precipitate heavy metals out of solution (Barber 1978). Such treatment has the dual advantage of precipitating the metal carbonate while holding pH within a narrow range at nearly optimum conditions. Although sodium bicarbonate is not as efficient in removing metal from solution as other bases, it has the advantage of neutralizing excess acidity, which helps in meeting wastewater discharge standards. The sodium bicarbonate acts as a buffer, maintaining alkalinity near the optimum pH level. Some metals, such as zinc, do not readily precipitate, regardless of the amount of carbonate added. By mixing soda ash (sodium carbonate), sodium bicarbonate, and lime (calcium hydroxide), however, it is possible to precipitate zinc as zinc hydroxide, while using the carbonates to stabilize pH. Sodium bicarbonate treatment has the additional advantage of easy handling, simple application, ability to function in continuous flow operation, and moderate cost (Barber 1978).

Patterson, Allen, and Scala (1977) studied the feasibility of carbonate precipitation in the removal of heavy metals. For nickel and zinc, no benefit was realized by using carbonate precipitation as opposed to hydroxide precipitation; the optimum pH for metal removal corresponded to pH values predicted by the theoretical metal hydroxide solubility diagram. No advantages in terms of denser sludges or better filtration characteristics were observed for the zinc carbonate or nickel carbonate systems. Beneficial results were observed using carbonate precipitation in cadmium and lead removal. Comparable residual cadmium concentrations were observed at approximately two pH units lower with carbonate treatment than with hydroxide treatment. The cadmium carbonate precipitates had approximately the same filtration rate as the cadmium hydroxide system. Treatment equivalent to that for lead hydroxide at pH 10.5 was obtained with the lead carbonate system at pH 7.5 and a total carbonate concentration of 0.08 moles/L, or at pH 10 and a total carbonate concentration of 0.002 moles/L. The lead carbonate system yielded a denser precipitate with improved filterability characteristics than did the lead hydroxide system.

Additional laboratory carbonate precipitation studies have been performed by McAnally, Benefield, and Reed (1984) and McFadden, Benefield, and Reed (1985), in which combined chemical treatment was performed for removal of nickel from solution. These studies are described in Subsection 3.4.5.

3.3.3 Sulfide Precipitation

Sulfide precipitation has been demonstrated to be an effective alternative to hydroxide precipitation for removal of heavy metals from industrial wastewaters (Bhattacharyya, Jumawan, and Grieves 1979; Bhattacharyya, Jumawan et al. 1981; Bhattacharyya, Shin et al. 1981; US EPA 1980; Kim 1981; Kim and Amodeo 1983; Ku 1982, 1986; Ku and Peters 1986, 1988; Peters et al. 1984; Peters, Ku, and Bhattacharyya 1984, 1985; Peters, Eriksen, and Ku 1985; Peters and Ku 1984, 1985, 1987, 1988). See table

3.5 (on page 3.34) for a listing of advantages and limitations of this technique. Avoiding sulfide reagent overdose prevents formation of the odor-causing H_2S. In currently operated soluble sulfide systems, which do not match demand, the process tanks must be enclosed and vacuum evacuated to minimize sulfide odor problems.

Bench-scale studies involving metal sulfide precipitation showed that sulfide precipitation is an extremely effective treatment in removing metals such as, Cd, Cu, Pb, Zn, As, and Se (Bhattacharyya, Jumawan, and Grieves 1979; Bhattacharyya, Jumawan et al. 1981; Bhattacharyya, Shin et al. 1981). Overall separation and precipitate settling rates were optimal for understoichiometric addition of sulfide (≈0.60 x the theoretical stoichiometric sulfide requirement) and pH >8.0 (Bhattacharyya, Jumawan, and Grieves 1979). Bhattacharyya, Shin et al. (1981) observed essentially complete removal of zinc using sulfide precipitation (1.0x) for pH >4. Measurements taken on H_2S loss (as gas) showed H_2S was negligible, attributable to the preference of the metal sulfide reaction over the H^+-S^{-2} reaction. Nickel precipitation with sulfide was found to be a strong function of reaction time for pH <10 in open systems. This was attributed to nickel dissolution causing the formation of $Ni(SOH)_2$ and $NiSO_4$ in the presence of oxygen.

The particle size distributions of ZnS, CdS, and NiS precipitation were studied, because removal efficiency and ease of removing the sludge are coupled. Both heavy metal removal and particle size distributions (PSDs) were reported (Ku 1986; Peters et al. 1984; Peters, Ku, and Bhattacharyya 1984; Peters and Ku 1987). Using sulfide precipitation, virtually all the zinc was removed at pH 8 to 10 (residual Zn <0.30 mg/L; removal efficiency >99.7%). Few particles greater than 20 μm in size were observed; the dominant particle size of the precipitates was only 5 to 7 μm, causing a cloudy appearance in the reactor suspension and making sedimentation and filtration operations difficult. This suggested that the use of a flocculant or coagulant aid would be advantageous. Although very high supersaturations were achieved, the kinetic order was low (i <1.02) for continuous ZnS precipitation (Peters et al. 1984). Little effect was observed on the resulting PSD by varying the reactor detention time. Increasing the suspended solids concentration increased both the particle growth rate and the precipitate dominant size, yielding a more favorable PSD. The presence of calcium improved the settling characteristics of both the Cd-Ca-Na_2S and the Cd-CaS slurry systems. The metal sulfide reactions were extremely rapid; chemical equilibrium was achieved within 5 minutes reaction time. A phase transformation from an amorphous, kinetically-favored precipitate to a more crystalline, thermodynamically-favored precipitate was indicated for the ZnS system as the precipitate aged. Patterson, Allen, and Scala (1977) and Patterson and Minear (1975) have also noted such phase transformations. Particle size, rather than completeness of the solid phase formation, often controls the apparent removal effectiveness.

The batch precipitation of zinc, cadmium, and nickel using both hydroxide and sulfide precipitation at various pH conditions, with and without complexing agents, has been studied (Peters and Ku 1985,1988; Ku and Peters 1988). Addition of EDTA inhibited zinc removal in both hydroxide and sulfide treatment because of the formation of stable metal chelates; the effect was more pronounced for the $Zn(OH)_2$ system (Peters and Ku 1985). Tartrate severely hindered both zinc hydroxide and zinc sulfide precipitation, resulting in the formation of very fine precipitates; this was confirmed when no zinc removal occurred even with a settling time of 30 minutes (Peters and Ku 1988). For removal of cadmium, virtually no change in residual cadmium concentration was observed in the presence of tartrate (compared with a control containing no complexing agents) because of the formation of low-stability complexes. Extremely low residual metal concentrations can be achieved using sulfide precipitation in the absence of chelating agents, as compared to similar hydroxide precipitation conditions (Peters and Ku 1985). The presence of phosphate enhanced the PSD because of a flocculation/agglomeration mechanism. The presence of ammonia has a minimal effect on metal sulfide removal and precipitation kinetics. The presence of EDTA severely inhibited CdS precipitation. Equilibrium conditions were reached quickly, within 5 minutes reaction time, for the ZnS and CdS systems (Peters and Ku 1985), while equilibrium was achieved after 40 minutes reaction time for the NiS system because of the oxidation of sulfide in the open system (Ku and Peters 1988).

Peters, Ku, and Bhattacharyya (1984) and Ku and Peters (1986) also addressed the effect of chelating agents on metal sulfide precipitation; EDTA forms very strong metal chelates that interfere with ZnS precipitation. Weak chelating agents, such as citrate, gluconic acid, and tartrate, form weak metal chelates; formation of the metal sulfide precipitate dominates in these cases. Removal of copper was nearly complete even in the presence of EDTA.

Hohman (1985) selectively precipitated Cd from a Cd-Fe wastewater at pH 2 by employing a stoichiometric sulfide dose only for the cadmium ions. At pH 6, Fe^{3+} was selectively removed from Cd through hydroxide precipitation. The CdS precipitation step goes to completion within 1 minute reaction time, while FeS precipitation is much slower. At high sulfide levels, FeS resolubilizes on formation of soluble species or colloids.

Laboratory studies were performed on a synthetic Pb-containing wastewater and a metal finishing plant effluent sample in Madras, India (Kamaraj, Jacob, and Srinivasan 1989; Kamaraj et al. 1990). Employing a soluble sulfide precipitation technique, removal of lead (in the absence of chelating agents) approached 100% for a sulfide dosage corresponding to 0.6x theoretical stoichiometric requirement. Removal exceeded 96% for pH in the range of 3 to 10. Ammonia had little effect on the removal of heavy metals from solution. Using 800 mg/L of EDTA, the removal of lead from the synthetic system was reduced to about 80%. Removal of Ni, Cr, Zn, Ag, and Cd exceeded 95% for pH >7. For sulfide dosages of 70% of the stoichiometric requirement, removal of the heavy metals exceeded 95%.

In a similar study, Kamaraj et al. (1991) investigated the effects of sulfide dosage, solution pH, and the presence of chelating agents (EDTA and cyanide) on treatment of a synthetic wastewater containing lead, cadmium, and silver. Stoichiometric sulfide dosage at pH 5, in the absence of chelating agents, resulted in maximum removals of Cd, Pb, and Ag (98, 73, and 100%, respectively). The maximum removal of Pb, Cd, and Ag occurred at pH 8, 4, and 4, respectively. To simulate an industrial wastewater, SO_4^{2-} was added to form a solution containing 1000 mg/L SO_4^{2-}, 40 mg Pb/L, 10 mg Cd/L, and 4 mg Ag/L. Maximum removal of lead, occurring at pH 8, was 95%, while maximum removal of Cd, occurring at pH 2 to 4, was 100%; similarly, maximum removal of Ag was 96% at pH 2. For a stoichiometric sulfide dosage at pH 8, the removal of Pb, Cd, and Ag with 100 mg/L EDTA present was 20, 82.4, and 72%, whereas with 10 mg/L cyanide (in place of the EDTA), the removal was 95.2, 96.8, and 64%. As a comparison, when a stoichiometric sulfide dosage was employed, the removal of Pb, Cd, and Ag was 90.4, 100, and 72%. When cyanide was present, the requirement for sulfide ions was increased to achieve maximum removal of all the heavy metals, except lead, in the presence of chelating agents.

There are two main processes for sulfide precipitation of heavy metals (US EPA 1980): soluble sulfide precipitation (SSP) and insoluble sulfide precipitation (ISP), the difference being the way in which the sulfide ion is introduced into the wastewater. This US EPA publication provides an excellent description and review of sulfide precipitation processes. In the SSP Process, sulfide is added to the wastewater in the form of a water-soluble sulfide reagent, such as sodium sulfide (Na_2S) or sodium hydrosulfide (NaHS). The addition of the solution may be monitored by periodic analyses of metal contents or it may be controlled by means of a feedback control loop employing ion specific electrodes. The process can be operated either in batch or continuous mode.

In the ISP Process, a slightly soluble ferrous sulfide (FeS) slurry is added to the wastewater, supplying the sulfide ions required to precipitate the heavy metals. Since most of the heavy metals are less soluble than ferrous sulfide, they will precipitate as metal sulfides. Since the FeS has a very low solubility with a sulfide concentration of 0.02 μg/L, emission of H_2S is minimized. In practice, FeS is freshly prepared by mixing $FeSO_4$ and NaHS. Among the advantages of the ISP Process is the absence of any detectable H_2S gas and reduction of Cr^{+6} to Cr^{+3}. Among its disadvantages is the considerably larger stoichiometric reagent consumption and generation of large quantities of sludge because of the ferrous hydroxide formation (US EPA 1980). The addition of FeS is not automatically controlled in response to metals content. The rate of FeS addition is determined by jar tests on the wastewater before it enters the sulfide precipitation tank. The process normally requires 2 to 4 times the stoichiometric amount of FeS (US EPA 1980). The use of an excessive amount of FeS adds to the chemical cost of the process; it also contributes to the production of large amounts of sludge. The Sulfex Process (Scott 1979) produces almost three times more sludge than the conventional hydroxide precipitation process (Kim 1981).

In the SSP Process, the high sulfide concentration often causes rapid precipitation of metal sulfides (high nucleation rates) resulting in small particulate fines and hydrated colloidal particles which have poor settling characteristics and poor filterability. In the presence of chelating agents, hydroxide precipitation is not possible, even at high pH. With sulfide precipitation, heavy metal removal is possible even with chelating agents present, although the metal sulfide precipitation is influenced by the presence of chelating agents (Ku and Peters 1986, 1988; Peters and Ku 1984, 1985, 1987, 1988; Peters, Ku, and Bhattacharyya 1984). In the absence of chelating agents, little metal hydroxide precipitation occurs when the pH is below 6. Metal sulfide precipitation can be conducted over a very wide pH range, typically from pH 2 to pH 12 (see figure 3.9 on page 3.33). Because metal sulfides are less soluble than the corresponding metal hydroxides, better removal efficiencies are achieved over a broad pH range. In addition, metal sulfides are less amphoteric than the corresponding metal hydroxides and are, therefore, less likely to resolubilize. Metal sulfide sludges usually have smaller volumes and are easier to dewater than metal hydroxide sludges.

An alternative to using FeS in the ISP Process involves addition of calcium sulfide (CaS) (Kim 1981; Kim and Amodeo 1983). Under this technique, the problems of delivery of the insoluble sulfide ions and the generation of large amounts of sludge can be minimized through addition of CaS. The addition of CaS as a slurry produces easily settleable precipitates. Calcium sulfide particles act as nuclei for production of metal sulfide particles, and the dissolved calcium ions function as a coagulant. Since calcium is mostly dissolved in the wastewater after reaction, the increase in the sludge volume is minimal. For this same reason, the CaS dosage requirement is nearly stoichiometric (in contrast to the overstoichiometric dosage requirement for FeS).

Whang, Young, and Pressman (1981) designed a soluble sulfide precipitation system for the Tobyhanna Army Depot in Tobyhanna, Pennsylvania. The Depot operates an electroplating facility which discharged its wastewaters along with other wastewaters from the Depot to a trickling filter plant. The wastewaters from the electroplating facility included three primary streams:

- cyanide-bearing wastewater, containing cadmium, copper, and cyanide;
- chromium-bearing wastewater, containing sodium, chromate, and other additives; and
- other alkaline and acidic wastewater, containing sodium, nickel, aluminum, tin, iron, lead, zinc, chloride, sulfate, nitrate, phosphate, and other additives.

In addition to these streams, the plating facility also periodically dumped spent solutions (ranging from once per month to once per year). The flow from the facility averaged 67,500 L/day (18,000 gal/day), consisting of 7,500 L/day (2,000 gal/day) of chromium-bearing wastewaters, 15,000 L/

day (4,000 gal/day) of cyanide-bearing wastewaters, and 45,000 L/day (12,000 gal/day) of acid/alkaline wastewaters. This system was installed at the Depot and has been in use since late 1981.

A full-scale SSP plant (Resta et al. 1978) was constructed by the U.S. Army at the Belvoir Research and Development Center in Fort Belvoir, Virginia, and became operational in February, 1983. Safety features of the plant included neutralization of the wastewater pH prior to sulfide addition, automatic control of the sulfide feed via a specific ion probe, addition of ferrous sulfate to remove the excess sulfide, hydrogen peroxide oxidation of the residual effluent sulfide, and covering and ventilation of the process tanks. Removal exceeding 90% was observed for Cd, Cr, Cu, Ni, and Zn, with removal of Pb exceeding 80%. The sludge generated averaged 0.3 L of sludge/1,000 L of wastewater treated; the solids content of the filter cake averaged 23.4% without the use of any sludge conditioners. The sludge samples were determined to be nonhazardous by the Extraction Procedure Toxicity test. Chemical costs in 1978 averaged \$0.08/1,000 L (\$0.30/1,000 gal) of wastewater treated, causing the process operational cost to be \$0.19/ 1,000 L (\$0.71/1,000 gal) excluding manpower and energy costs.

Peters and Ku (1984) investigated the continuous precipitation of copper, nickel, chromium, and zinc from synthetic and actual industrial plating wastewaters by both hydroxide and sulfide treatment. For a zinc-nickel wastewater, slightly lower residual and enhanced floc size were observed in sulfide treatment than in hydroxide treatment. Low pH treatment ($pH \approx 7.2$ to 7.4) resulted in incomplete removal of nickel; zinc precipitation was preferential to that of nickel. At higher pH levels ($pH \approx 10$), removal of both nickel and zinc exceeded 98% through either hydroxide or sulfide treatment. Larger sulfide dosages and higher pH conditions resulted in greater metal removal and larger settling velocities. Chromium was not effectively removed from the Cu-Ni-Cr-Zn wastewater through hydroxide treatment. Increasing the sulfide dosage resulted in lower residual heavy metal concentrations for Cu, Cr, and Zn, while the nickel concentration increased because of the formation of fine flocs. At pH 10.0, in sulfide treatment, removal exceeded 97%, 98%, 55%, and 83% for Zn, Cu, Cr, and Ni, respectively. The understoichiometric addition of sulfide showed great promise as a treatment technique to achieve extremely low metal residuals, while minimizing the potential for H_2S gas evolution and sulfide toxicity problems.

In a more scientific investigation, Bhattacharyya, Jumawan et al. (1981) found sulfide precipitation using Na_2S to be highly effective in removing Cd, Cu, Pb, Zn, As, and Se from complex wastewaters. Full-scale plant data were obtained at the Boliden Metall Corporation's metal smelting plant in Skelleftehamn, Sweden. The full-scale plant was put into operation in 1978 and was designed to precipitate As, Cu, Cd, Hg, Pb, and Zn as sulfides at pH 3 to 5 by recycling to "roaster"; the fluoride was removed separately by lime treatment (at pH >10) forming CaF_2. The wastewater was first partially neutralized (to pH 2.5 to 3.0) with addition of NaOH, after which Na_2S was added (as a 15% Na_2S solution). The reagent dosages were controlled by monitoring pH. The sulfide precipitate was removed by sedimentation (after polymer addition) and post-filtration. The wastewaters treated also contained fluoride, which was removed by precipitation with lime (using a 10% lime slurry) following the metal sulfide precipitation. Five separate tests were performed to determine the extent of heavy metals and arsenic removal and to establish the performance of the individual vessels. The process had been in operation for nearly a year at the time of publication.

For this plant's wastewater, removal of Cd, Cu, and Zn exceeded 98%, with removal of As and Se being 98% and >92%. The residual concentrations achieved for Cd, Cu, and Zn were consistently in the range of 0.05 to 0.10 mg/L. When only hydroxide treatment was used, the settling rates and metal separations were consistently lower than those obtained by sulfide precipitation. The Cd, Zn, and Se removal was much poorer, even at pH 10.5, using hydroxide treatment. At pH 8.5, the residual metal concentrations of Se, Cd, and Zn were 9, 2, and 5 mg/L. At pH 10.5, the residual Cd and Zn concentrations decreased to 0.6 and 1.1 mg/L. The settling rate of the sludges was a function of pH and sulfide dosage. Poor settling velocities resulted from hydroxide precipitation and overstoichiometric sulfide precipitation. The settling rate, at 0.6x the stoichiometric requirement of sulfide, was twice that obtained using hydroxide precipitation. This combined hydroxide-sulfide treatment at pH 8 to 9 was shown to be effective for removal of Cd, Cu, Hg, Fe, Pb, Zn, and As in both synthetic wastewater and for full-scale treatment of wastewater. Even with sulfide overdoses at low pH, no H_2S gas loss was observed.

3.3.4 Xanthate Precipitation

In xanthate treatment, metal contaminants exchange with sodium ions contained in the xanthated material to form an insoluble complex. The xanthate acts as an ion exchange material, removing heavy metals and replacing them with sodium and magnesium. The heavy metals-laden material can be removed from solution by sedimentation and filtration. Compared with metal hydroxide precipitation, xanthate treatment offers the following advantages (Federal Remediation Technologies Roundtable 1992):

- a higher degree of metal removal;
- less sensitivity to fluctuations in pH (metal xanthates do not exhibit amphoteric solubilities);
- less sensitivity to the presence of complexing agents;
- improved sludge dewatering properties; and
- the capability of selectively removing metals.

Wing and Rayford (1977) state that the process will probably not be economical for initial metal concentrations exceeding 100 mg/L, although xanthate treatment could be used as a secondary treatment to further lower the metal concentration to below discharge limits.

The process was developed by the U.S. Department of Agriculture (Wing, Doane, and Russel 1975; Wing et al. 1978; Wing 1974; Wing and Rayford 1975, 1976). Xanthates are sulfonated organic compounds. The xanthate-metal precipitation process can be represented as follows:

$$ROCSSNa + M^{+} \xrightarrow{NaOH} ROCSS - M + Na^{+} \qquad [3.1]$$

or

$$2(ROCSSNa) + M^{2+} \xrightarrow{NaOH} ROCSS - M - SSOCR + 2Na^{+} \qquad [3.2]$$

where M^+ or M^{2+} are the metal ions and NaOH indicates that the reaction occurs at high pH.

Whereas hydroxide precipitation is effective over a pH range of approximately 9 to 12, xanthate precipitation is effective over a much wider pH range ($\approx$ 3 to 12), with maximum effectiveness above pH 7. Solutions with pH less than 3 rapidly decompose the xanthates (Wing 1974). The hierarchy for selective removal of some cations and heavy metals by xanthate treatment is in the following order: Na << Ca-Mg-Mn < Zn < Ni < Cd < Pb-Cu-Hg (Flynn, Carnahan, and Lindstrom 1980). This technique still produces significant quantities of sludge which must be handled in accordance with RCRA and other applicable regulations.

Wing (1974) observed that contaminants could be introduced when water was treated with starch xanthates and a cationic polymer; possible contaminants include small ionic species (Cl^- from the cationic polymer, and Na^+, OH^-, and CO_3^{2-} from the xanthate), small nonionic species (CS_2 and COS from the xanthate), and the polyelectrolytes themselves. The presence of sequestrants (diglycolate, NTA, polyphosphate, or citrate were used in this study) at concentration levels of 0.1 g/L did not affect efficiency in removing mercury. Starch xanthate effectively treated 11 other metals: Cd^{2+}, Cr^{3+}, Cu^{2+}, Fe^{2+}, Fe^{3+}, Pb^{2+}, Mn^{2+}, Hg^{2+}, Ni^{2+}, Ag^+, and Zn^{2+}. The treatment can be performed using either batch or continuous precipitation. Wing and Rayford (1976) reported that the insoluble starch xanthate-metal sludge settled rapidly and dewatered to 50 to 90% solids content after filtration or centrifugation. A preliminary estimate of the cost to make the insoluble starch xanthate was \$0.68/kg (\$0.30/lb). Other starch-based products ranged in price from \$0.68 to \$1.69/kg (\$0.30 to \$0.75/lb). The chemical cost of treating a 50 mg/L Cu-EDTA rinse with lime and polymer was estimated to be \$0.02/1,000 L (\$0.07/1,000 gal) (Wing and Rayford 1976).

Wing and Rayford (1977) conducted bench-scale studies on synthetic wastewater that contained heavy metals using insoluble starch xanthate treatment. The heavy metals investigated included Ag^+, Au^{3+}, Cd^{2+}, Co^{2+}, Cr^{3+}, Cu^{2+}, Fe^{2+}, Hg^{2+}, Mn^{2+}, Ni^{2+}, Pb^{2+}, and Zn^{2+}. The initial heavy metal concentrations ranged from 26 to 104 mg/L. The residual metal concentration was generally below the effluent discharge standards. Wing and Rayford (1976) showed that other starch-based products were effective in removing heavy metal; these products were insoluble starch xanthate, carboxyl cross-linked starch, polyethylenimine cross-linked starch, and tertiary amine and quaternary ammonium cross-linked starch.

Wing (1974) conducted bench-scale studies on a synthetic mercury-containing wastewater in the presence of various sequestrants (diglycolate, NTA, polyphosphate, citrate) and with a control containing no sequestrants. The presence of the sequestrants at a concentration level of 0.1 g/L did not adversely affect the removal of mercury. Treatment of the mercury solution having initial pH in the range of 3 to 11 was effective for removal of Hg. Starch xanthate and cationic polymer could be added in either order for effective removal of Hg. The more slowly the starch xanthate was added to the solution, the more effectively it complexed with mercury, indicating that the complexation was not an instantaneous reaction. Slower precipitation generally produces fewer and larger particles, resulting in a sludge which is more easily filtered. Reaction times required were on the order of five minutes. When a larger bench-scale system (95 L (25 gal)) was used, a solution containing 31,770 μg/L Cu^{2+} was reduced to a residual level of 22 μg/L Cu^{2+} with a 5 minute contact time, and to a level of 20 μg/L with a 120 minute contact time.

In a laboratory investigation, Bricka and Cullinane (1987) prepared several synthetic sludges containing the contaminant cations Cd, Cr, Hg, and Ni. The wastes were treated through hydroxide and xanthate (cellulose and starch) precipitation. The treated sludges were subjected to Extraction Procedure (EP) Toxicity tests. All the solidified sludges passed the EP Toxicity test, except for the solidified hydroxide sludge, which failed the test for Hg. The unsolidified cellulose xanthate sludge failed the EP Toxicity test for Ni and Cd, while the unsolidified hydroxide sludge failed for every metal tested. The xanthate-precipitated sludges appear to effectively immobilize heavy metals.

Fender, MacGregor, and Patterson (1982) studied sulfide precipitation of zinc-laden foundry wastewaters. The wastewater pH was adjusted with lime to provide pH ranging from 8.5 to 11.0. Ferrous sulfide (at a dosage of 750 mg/L) was added to each sample. The residual lead and iron concentrations were consistently less than 0.1 mg/L when initial metal concentrations in the wastewater were 21.1 and 0.1 to 1.0 mg/L, respectively. The zinc concentration decreased from 775 mg/L to a level of 1.7 to 3.7 mg/L; however, this removal was still inadequate. Two-stage, hydroxide-sulfide precipitation was investigated in pilot-scale treatability studies. The best treatment occurred with lime at pH 9.6 with 20 mg/L FeS added to the supernatant. This resulted in a final filtered effluent concentration of 0.05 mg Zn/L.

3.3.5 Combined Precipitation Treatment

In a very broad sense, each precipitation system (with the exception of metal hydroxide precipitation) involves a combined precipitation system. This is because the precipitations are generally performed at a particular pH. For pH >6, metal hydroxide precipitation is possible. Metals are preferentially removed from solution by sulfide precipitation. However, coprecipitation of metal sulfides and metal hydroxides is possible. The majority of investigations involving combined precipitation systems have been bench-scale studies.

McAnally, Benefield, and Reed (1984) studied the effectiveness of soluble sulfide and carbonate in reducing nickel in a synthetic nickel-plating wastewater. Employing jar tests, the investigators determined optimum pH range for nickel removal from the synthetic wastewater to be 10.0 to 11.0. Optimum nickel removal occurred at pH 11 where a residual total nickel concentration of 0.1 mg/L was obtained with a sulfide:nickel weight ratio of 2.0 and a carbonate:nickel weight ratio of 20.0. At pH 10, a similar degree of removal (0.2 mg/L residual total Ni) was obtained using a CO_3^{2-}: Ni^{2+} ratio of 10.0 and a S^{2-}: Ni^{2+} ratio of 0.5. The excellent results in removing nickel were probably due to a coprecipitation phenomenon. To treat the synthetic nickel wastewater, the pH was adjusted by drop-by-drop addition of 1N NaOH, and an equivalent amount of 1N $CaCl_2$ solution was added to simulate lime addition. The carbonate was added in the form of $NaHCO_3$. Such conditions likely led to the precipitation of calcium carbonate ($CaCO_3$), which has been shown to be an excellent adsorbent for Cd, Pb, and Zn (Chang and Peters 1985; Faust and Schultz 1983; Peters and Chang 1984, 1985). Coprecipitation and adsorption of $Ni(OH)_2$ and NiS onto the $CaCO_3$ surfaces may have caused the excellent results in removing Ni. However, McAnally, Benefield, and Reed (1984) did not report the residual calcium concentrations needed to confirm this supposition.

In a similar study, McFadden, Benefield, and Reed (1985) investigated the effect of iron as a coprecipitator of nickel as well as the effects of carbonate addition, pH adjustment, and polymer addition. For pH adjustment, along with carbonate addition, optimum nickel removal occurred when the total carbonate concentration (C_T) was 50 mg/L at pH 11. These conditions resulted in the soluble and total nickel concentrations of <0.10 and 0.10 mg/L. All three C_T concentrations, employing (50, 100, and 200 mg/L) at pH 10 and 11, effected removal of at least 96% of the total nickel and 99% of the soluble nickel. The investigators suggested that the calcium may provide a nucleus for $CaCO_3$ formation, thereby increasing settleability, although they do not report the final residual calcium concentrations of the treated wastewater. The initial Ca^{2+} concentrations were the same as those employed by McAnally, Benefield, and Reed (1984). Coprecipitation of Ni onto the $CaCO_3$ surfaces may indeed be an explanation for the high percentages of nickel removal observed.

Nickel removal through hydroxide precipitation was the most efficient for the synthetic wastewater at pH 10 to 11, depending on the Fe:Ni ratio and C_T (McFadden, Benefield, and Reed 1985). Both the soluble and total nickel concentrations at pH 10 (Fe:Ni = 2, and $C_T = 0$), were reduced to <0.10 mg/L. Identical results ensued at pH 11 (Fe:Ni = 2, and $C_T = 100$ mg/L as $CaCO_3$). At pH 9, the best overall removal was effected with a Fe:Ni ratio of 1.0 and $C_T = 50$ mg/L as $CaCO_3$, where the total and soluble residual nickel concentrations were 0.20 and 0.10 mg/L. For the actual wastewater, the most efficient soluble nickel removal occurred at pH 10 with a Fe:Ni ratio of 0.7 and $C_T = 0$, resulting in total and soluble nickel concentrations of 0.30 and 0.25 mg/L, respectively. Use of anionic and cationic polymers did not enhance the removal of nickel appreciably. For the actual wastewater, the lowest cost to treat the wastewater was $ 0.14/1,000 L ($ 0.53/1,000 gal) (in 1985 dollars) for conditions of pH 10, Fe:Ni ratio of 0.7, and $C_T = 0$.

When ferrous sulfide was used as a coprecipitator, heavy metals (Cu, Cd, Ni, Cr, and Zn) in the influent wastewater were shown to be significantly reduced in concentration (Schlauch and Epstein 1977). FeS treatment was found to be superior to conventional hydroxide precipitation employing lime as the precipitant.

Chang and Peters (1985) observed that cadmium could be very effectively removed through conventional lime softening operations; the maximum contaminant level of 0.01 mg/L for Cd could be met with pH in the range of 7.3 to 11.0. Calcite was the only morphological form observed in the continuous $CaCO_3$ precipitation. The residual calcium concentration increased ≈ 30 to 40 mg/L in the presence of cadmium, indicating an inhibitory effect on $CaCO_3$ precipitation. Removal of cadmium was attributed primarily to physical adsorption onto the $CaCO_3$ sludges.

Talbot (1984) described a process using less than stoichiometric addition of sulfide, giving a combined hydroxide-sulfide treatment. At pH 8.0, the

Talbot Process reduced the cadmium concentration in a solution from 15.0 mg/L to <0.05 mg/L, while hydroxide treatment provided a residual concentration of 4.8 mg/L. For a water containing 2.9 mg/L Hg, conventional hydroxide treatment did not remove any mercury, while the Talbot Process lowered the mercury level to <0.001 mg/L at pH 8.0. The operating cost of the Talbot Process is comparable to that of conventional hydroxide precipitation. A smaller quantity of sludge is generated by the Talbot Process (compared with conventional hydroxide precipitation), thereby lowering sludge disposal costs.

The Talbot Process basically involves the understoichiometric addition of sulfide to the wastewater. Bhattacharyya, Jumawan, and Grieves (1979) also observed that adding 60% of the theoretical requirement of sulfide provided effective removal of heavy metals (Cu, Cd, Hg, Pb, and As). Peters, Eriksen, and Ku (1985) also observed that understoichiometric addition of sulfide, even in an amount as low as 0.5 x stoichiometric requirement, likewise provided excellent removal of zinc and cadmium and decreased the resulting sludge volume. They suggested that metals could be selectively precipitated, removed, and recovered from a mixed-metal wastewater by properly controlling the solution pH, sulfide dosage, chelant dosage, type of chelant, and temperature in a cascading series of reactors. For example, zinc can be selectively precipitated at pH 6.0, with stoichiometric sulfide dosage and addition of EDTA (Peters, Eriksen, and Ku 1985). Pugsley et al. (1970) were the first investigators to observe that selective removal of a particular heavy metal could be effected in a cascading reactor system through proper control of the dosage rates. A preliminary estimate for the sulfide treatment of plating wastewater was approximately $2.45/1000 L ($9.27/1,000 gal) which compared favorably with the $2.50/1,000 L ($9.45/1,000 gal) approximate cost of treating it by conventional hydroxide precipitation (Peters, Eriksen, and Ku 1985). Considerable cost savings can be realized also through reuse, recycle, and recovery of the waste metals from the plating process.

Higgins and Slater (1984) observed that treating a mixture of metals is a somewhat more effective measure than treating metals individually. Sulfide treatment lowered the solubility of nickel and cadmium. Ferric hydroxide precipitates reduced the metal solubilities through the incorporation of other metals into an amorphous precipitate and the provision of surface sites for adsorption. Addition of ferrous sulfate to an alkaline environment (pH between 7 and 10) caused the iron to precipitate as iron hydroxide and aided in the flocculation of solids in the process. Such treatment was very effective in reducing hexavalent chromium to the trivalent form. A combination of ferrous sulfate and sodium sulfide produced a sludge that was easily removed, yet minimized sludge production.

Pilot-plant studies (Maruyama, Hannah, and Cohen 1975) employing lime addition (260 mg/L giving rise to a pH of 10.0) plus 20 mg/L ferrous sulfate were used to treat a nickel wastewater initially containing 5 mg/L Ni. The residual nickel concentration was reduced to 0.35 mg/L after sedimentation and filtration. Hydroxide precipitation (at pH 11.5, $Ca(OH)_2$ dosage = 600 mg/L) resulted in a residual nickel concentration of 0.15 mg/L after sedimentation and filtration. Leckie, Merril, and Chow (1985) likewise observed that trace elements of cadmium, zinc, lead, arsenic, selenium, silver, chromium, copper, and vanadium could be removed by adsorption/coprecipitation with amorphous iron oxyhydroxide. In both anion and cation adsorption, iron dose and solution pH were the two adsorption controlling parameters.

Of the three precipitation processes, carbonate precipitation produced the smallest sludge volume. Neither the hydroxide sludges nor the sulfide sludges thickened well in the clarifier. When subjected to acidification, the hydroxide sludge was the least stable, while the sulfide sludge was the most stable.

Brantner and Cichon (1981) compared the use of hydroxide, carbonate, and sulfide treatments in removing heavy metals (Zn, Cr, Cd, Cu, and Pb). Effective zinc removal was achieved with all three chemical precipitate processes. For low soluble zinc levels in the clarifier overflow, it appeared that zinc removal was not limited by solubility, but rather by the effectiveness of the separation of solids and liquid. The inability of carbonate precipitation to consistently effect low residual zinc concentrations was attributed to the kinetics of zinc carbonate formation, causing the solubility of zinc hydroxide to govern the removal of zinc. Chromium was effectively removed only through sulfide precipitation. Precipitation of cadmium and copper was very effectively achieved through all three processes. Lead was effectively removed through carbonate and sulfide precipitation, causing a residual filtered lead concentration of <0.1 mg/L; hydroxide precipitation resulted in a mean effluent level of 0.2 mg/L.

3.4 Scientific Basis

3.4.1 Substitution Processes

In the early 1980s, several groups of investigators (Brunelle and Singleton 1983; Rogers 1983; Klee, Rogers, and Tiernan 1984; Kornel and Rogers 1987; Franklin Research Center 1982; Neuman and Sasson 1983; Friedman and Halpern 1992b) described the reaction of PCBs with the phase transfer catalyst, PEG, in the presence of KOH. The destruction of the halogenated carbon compounds by PEG in the presence of a base, such as KOH, was found to be a promising technique for pollution control.

For example, Brunelle and Singleton (1983) described the use of PEG and poly(ethylene glycol) methylethers (PEGMs) and their reactions as nucleophiles with PCBs and similar chlorinated aromatics under basic and mild reaction conditions. In this reaction, partial dechlorination of PCBs occurred with PEG, producing aryl polyglycols in two hours at 100°C (212°F). The rate of reaction of PCBs with KOH/PEG was faster than that with KOH/PEGM; KOH is more effective than NaOH as a base when used with an equivalent amount of PEG. Possibly, PEG functions as a phase transfer agent, complexing the potassium ion and transporting it into the organic phase (Bailey and Koleske 1976; Yanagida, Takahashi, and Okahara 1977). An alternative possibility is that the PCB is extracted into the glycol phase followed by the reaction in the polar glycol phase. Efficient stirring is important to the success of the reaction.

The Franklin Institute (Franklin Research Center 1982) developed a related process. They found that PCBs could be destroyed by reaction with sodium metal and PEG at 165 to 180°C (329 to 356°F) and theorized that sodium PEG was the active reagent. Brunelle and Singleton (1983) compared the two methods. The investigators used a 10-fold excess of NaPEG in a solventless reaction with Aroclor 1260, and 60% reaction occurred after two hours at 75°C (170°F) under nitrogen. Air bubbled through the reaction mixture, giving only 23% reaction. A 5-fold excess of KOH/PEG under the same conditions resulted in 100% destruction of PCBs. In this study, PEGM provided better detoxification than PEG and a "no water" situation was better than when 15% water was present.

The use of PEGMs as phase transfer catalysts (Brunelle and Singleton 1983) in the alkoxylation of mono- and dichlorobenzene has been reported by Neuman and Sasson (1983). Using polyethylene glycol of mean molecular weight of 6000, (PEG-6000) with n-pentanol in addition to the basic hydroxide with o- and p-dichlorobenzenes, the investigators produced corresponding ortho- and para- pentoxy chlorobenzene and other products. Differences in the affinity of tertiary and secondary alkoxides for the glycol catalyst were noted when compared with that of the primary type. The method of Brunelle, Franklin Research Center, Kornel, Rogers, and Sparks, and that of Neuman involved reaction of PEG with alkali metal hydroxides, displacing the resistant chloride in the aromatic chloride via nucleophilic displacement and, possibly, holding fast the remaining residue in the water soluble and/or in organic solvents. The reaction mechanism could be either an addition-elimination sequence (Macomber et al. 1983) or an S_{RN1} mechanism (Bunnett 1978).

The S_{RN1} mechanism involves a chain reaction with electron transfer to the aromatic nucleus followed by expulsion of the chloride ion (Cl^-) and combination with stabilization of the aromatic radical nucleus. The electron-donating nucleophile should have a low ionization potential in order to accommodate the aromatic rings' electron affinity. It is a radical chain mechanism with nucleophilic substitution. Equation [3.6], below, is the summation of the preceding equations, where X^- is a chloride (Macomber et al. 1983).

$$[RX]^- \rightarrow R\bullet + X^- \qquad [3.3]$$

$$R\bullet + Y^- \rightarrow [RY]^- \qquad [3.4]$$

$$[RY]^- + RX \rightarrow RY + [RX]^- \qquad [3.5]$$

$$RX + Y^- \rightarrow RY + X^- \qquad [3.6]$$

In order for chlorinated aromatics to undergo the above addition-eliminating sequence, the aromatic halide would be required to have electron withdrawing groups (multiple of 4, 5, or 6 chlorides) substitution in order to activate the aromatic ring for the above addition-elimination mechanism.

Acyclic PEG shows complexing properties of ether oxygens with metals in a manner similar to those of crown-ethers (Balasubramanian and Chandani 1983). PEG with a general structure $HO(CH_2CH_2\text{-}O)_nH$ differs from crown-ethers essentially in that the macrocyclic ring is opened and the molecule is laid out. In the solution state, hydrodynamic evidence (Bailey and Koleske 1976) indicates that PEG chains are randomly coiled or in an unordered conformation in several solvents. Yanagida, Takahashi, and Okahara (1977) have shown that linear PEG can complex several alkali salts and can effect phase transfer into organic solvents with efficiencies comparable to those obtained with 18-crown-6. Complexation does not seem to occur when PEG of molecular weight of less than 300 (PEG-200) is used. This suggests that more than six ethylene oxide moieties are necessary for complexation (Balasubramanian and Chandani 1983).

Laboratory, pilot-scale, and full-scale data on the use of substitution reactions to treat contaminated materials are available for the following media and contaminants:

SOILS

- PCBs, dioxins, PCP

DEBRIS

- cyanides

WATER

- cyanides
- phenols
- metals (precipitation)

OTHER

- PCBs and dioxins in mineral oil

Chemical treatment is rarely used upon high concentrations of target compounds in any medium usually because of the large amounts of reagent normally required and, consequent high reagent costs. Another difficulty, because some of the chemical reactions are very energetic, lies in controlling them at high concentrations. Note that this is not a firm rule. For example, KPEG reagents have been used to commercially treat small quantities of high concentration PCB waste at 0.1 to 10% of chlorinated biphenyls.

3.4.2 Oxidation Processes

This section on oxidation processes focuses primarily on organic destruction and assesses those technologies that appear to be both innovative and feasible for application to multiphase pollution in the near future. Such traditional oxidants as chlorine or permanganate are not addressed because they are not innovative and are not appropriate for use at many hazardous waste sites. This discussion is centered mostly on excited state and free radical processes involving various activated oxygen species; primary among them is the hydroxyl free radical, OH•. These species are common, short-lived, kinetic intermediates in a wide range of oxidation reactions, including enzymatic and combustion reactions.

Excited states are energetically rich intermediates produced by photochemical processes involving molecular absorption of light or photons. Excited state species exhibit a reactivity generally much higher than their ground state analogues and hence, are inherently unstable. Cleavage of chemical bonds generates free radical molecular fragments, which bear unpaired electrons. They exhibit significantly higher reactivities than the parent molecule and, like excited states, are inherently unstable. Free radicals and excited states are formed with energy absorption by a substrate. An example of this is the photolysis or radiolysis of water (Singh 1986):

$$H_2O \xrightarrow[Vac-UV\ photolysis]{\gamma-\ or\ e^- -irradiation} H\bullet + OH\bullet + H_2O^+ + e_{aq}^- + H_2O* \quad [3.7]$$

where H_2O* represents an excited state species.

Because of their high reactivities, these species undergo further reactions to produce other radicals or reactive species:

$$H_2O^* \rightarrow H\bullet + OH\bullet \quad [3.8]$$

$$OH\bullet + OH\bullet \rightarrow H_2O_2 \quad [3.9]$$

$$H_2O^+ + H_2O \rightarrow H_3O^+ + OH\bullet \quad [3.10]$$

This sequence of radical formation (initiation), successive radical formation (propagation), and ultimate return to stable species (termination) is called a chain reaction (Lowry and Richardson 1981). Chain reactions have the central advantage of being energy-efficient. At sufficiently high temperatures, most chemical bonds will break, forming radicals. At lower temperatures, less than 200°C (400°F), an initiator that produces radicals easily under milder conditions is needed.

3.4.2.1 Photolysis

According to quantum theory, light is composed of discrete particles called quanta or photons that carry an amount of energy ε determined by its wavelength (λ) or its frequency (υ) according to:

$$\varepsilon = hc / \lambda = h\upsilon \qquad [3.11]$$

where h is Planck's constant and c is the speed of light in a vacuum. Atoms and molecules can absorb a quantum only when the energy of the quantum matches an energy transition to a higher state (Lowry and Richardson 1981). A single quantum can bring about a single transition only if it is adequately and specifically energetic (proper wavelength). The intensity of light, on the other hand, influences the number of molecules able to undergo a transition, but does not change the nature of the transition.

Different wavelengths of light produce different kinds of molecular transitions. Higher vibrational or rotational levels are induced by infrared excitation. High-energy ultraviolet radiation can cause ionization by the removal of an electron from the molecule. Longer wavelength UV and visible light may promote a molecule from its ground state to an excited state.

One mole of photons is called an Einstein, and the energy of an Einstein of light at a certain wavelength can be calculated. For example, the energy contained in an Einstein of 200 nm light is 143 kcal, as opposed to that of 700 nm light which is only 40 kcal. The bond energies of C-C and C-H are approximately 100 kcal/mole, which is equivalent to the energy contained in an Einstein of 286 nm light (Lowry and Richardson 1981). Saturated hydrocarbons absorb light only below 200 nm and their photochemistry is not very interesting, because in their excited states these molecules fragment indiscriminately through sigma bond cleavage. Molecules having unsaturated bonds and/or hetero atoms such as N or O are more photochemically active. With increasing conjugation there is a decrease in the energy of the lowest transition.

Although there may be many transitions possible for an excited state, not all transitions are equally probable. The extinction coefficient, ε, of a compound is an experimentally determined parameter indicating the probability of a certain transition at a certain wavelength. The greater the extinction coefficient, the more favorable the transition and the more strongly the compound will absorb at that wavelength.

3.4.2.2 UV-Hydrogen Peroxide

For certain organic compounds, UV light itself is capable of initiating bond cleavage processes, leading to degradation. The range of compounds that undergo degradation in the presence of UV light alone, however, is somewhat limited. The rate of UV-initiated degradation can be slow, and the extent of degradation can be incomplete. In contrast, the application of a combination of ultraviolet light and hydrogen peroxide in aqueous media produces hydroxyl radicals (OH•). Hydroxyl radicals are potent oxidizing agents that rapidly attack a wide range of organic materials and are capable of completely oxidizing dissolved organic contaminants in water to carbon dioxide, water, and salts (Heeks, Smith, and Perry 1991).

The process initially involves the UV-catalyzed decomposition of hydrogen peroxide into hydroxyl radicals:

$$H_2O_2 \xrightarrow{UV} 2OH\bullet \qquad [3.12]$$

As in any photochemically initiated process, the overall efficiency of the generation of products (in this case, hydroxyl radicals) is a function of the quantum yield (Φ) process:

Quantum Yield (Φ) = (Moles Substrate Reacted)/(Moles Photons Absorbed):

$$\Phi = \frac{(Moles\ \ Substrate\ \ Reacted)}{(Moles\ \ Photons\ \ Absorbed)} \qquad [3.13]$$

The quantum yield of photochemical reactions depends upon many factors including the optical path length of medium, the molar extinction coefficient and concentration of the substrate, and the intensity and wavelength of light source employed. The optical path length of the medium should be as large as possible to provide maximum light flux in optimizing the quantum yield of the process. Suspended solids, colloids, and particulates that will reflect and absorb light should be minimized. Large molar extinction coefficients in the substrate being irradiated will lead to greater quantum yields. The molar extinction coefficient of any substrate is dependent upon the wavelength of light with which it is irradiated. For hydrogen peroxide, the molar extinction coefficient at 254 nm is approximately 19.6 $M^{-1}cm^{-1}$ (Smith 1988). Consideration of these factors is essential in order to select the optimum light source to be used in the process. Once generated, the highly reactive hydroxyl radicals degrade organic materials in much the same way as hydroxyl radicals generated by other means.

3.4.2.3 Ozonation and Advanced Oxidation Processes

There are two modes of action in ozonation processes, illustrated in figure 3.10 and figure 3.11 (on page 3.59). That of figure 3.10 is along a slow

Figure 3.10
Free-Radical Chain Reaction of Ozone Decomposition

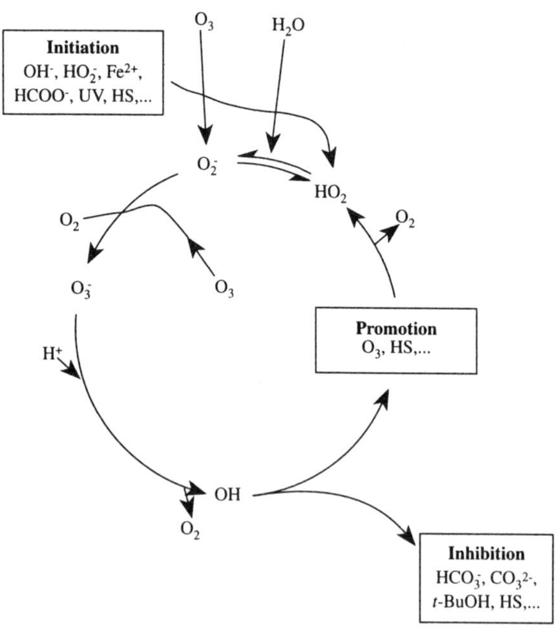

From Bablon, G., Bellamy, W.D., Bourbigot, M., Daniel, F.B., Dore, M., Erb, F., Gordon, G., Langlais, B., Laplanche, A., Legube, B., Martin, G., Masschelein, W.J., Pacey, G., Reckhow, D.A., Ventresque, C., *Fundamental Aspects*, 11, 17, 21, in *Ozone in Water Treatment: Application and Engineering*, Langlais, B., Reckhow, D.A., Brink, D.R., Eds., Lewis Publishers, a subsidiary of CRC Press, Boca Raton, Florida, 1991. With Permission.

Figure 3.11
Ozone Decomposition Catalyzed by Hydrogen Peroxide

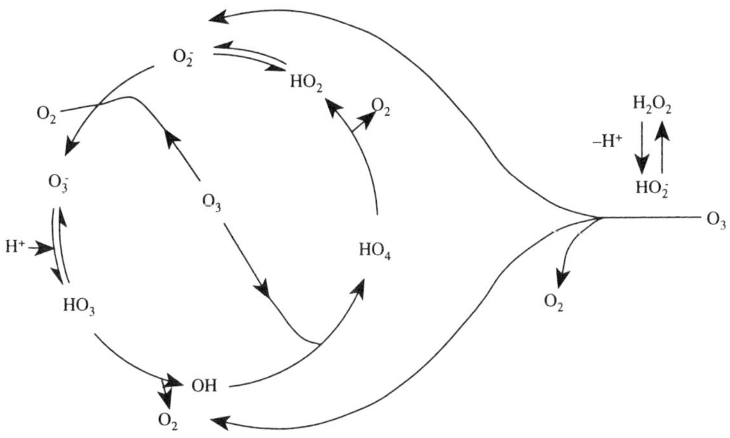

From Bablon, G., Bellamy, W.D., Bourbigot, M., Daniel, F.B., Dore, M., Erb, F., Gordon, G., Langlais, B., Laplanche, A., Legube, B., Martin, G., Masschelein, W.J., Pacey, G., Reckhow, D.A., Ventresque, C., *Fundamental Aspects*, 11, 17, 21, in *Ozone in Water Treatment: Application and Engineering*, Langlais, B., Reckhow, D.A., Brink, D.R., Eds., Lewis Publishers, a subsidiary of CRC Press, Boca Raton, Florida, 1991. With Permission.

reaction route that involves direct interaction with molecular ozone. This is a highly selective pathway and often does not result in complete mineralization of contaminants. The pathway depicted in figure 3.11 entails ozone decomposition in a series of reactions producing various free radicals, the most reactive among them being the hydroxyl radical (OH•). This reaction is initiated, and its rate is limited by, ozone reaction with hydroxide. In order to enhance the kinetics and energetics of OH• generation, ozone decomposition can be catalyzed by the addition of hydrogen peroxide and/or simultaneous use of UV irradiation (see figures 3.10 and 3.11). The resulting radical is highly reactive and attacks a wide variety of organics, oxidizing them to mineralization.

3.4.3 Precipitation Processes

Metals exist in aqueous systems in a variety of forms, including soluble, insoluble, inorganic, metal-organic complexes, reduced, oxidized, free metal, precipitated, adsorbed, and complexed. In any solid-liquid suspension, an equilibrium exists between the free ions of a given metal, the soluble ligands, and the solid phase. Soluble ligands may be inorganic or organic in nature, whereas the solid phase may consist of sludge plus metal precipitates. Treatment processes must be selected to remove the existing form of the metal. In general, there are five categories of metal species in aqueous systems: (1) free ions, (2) metal-hydroxyl complexes, (3) metal-ligand complexes, (4) adsorbed metals, and (5) the solid phase. These metal speciation categories are summarized below:

(1) Free ions, M^{+n}

(2) Metal-hydroxyl complex

$$M^{+n} + j\ OH^{-} <===> M(OH)_j^{\,n-j}; \qquad K_j = \frac{\left[M(OH)_j^{\,n-j}\right]}{\left[M^{+n}\right]\left[OH^{-}\right]^j} \qquad [3.14]$$

(3) Metal-ligand complex

$$M^{+n} + iL^{-m} <===> ML_i^{\,n-im}; \qquad K_i = \frac{\left[ML_i^{\,n-im}\right]}{\left[M^{+n}\right]\left[L^{-m}\right]^i} \qquad [3.15]$$

(4) Adsorbed metal

$$M + S <===> MS; \qquad K_A = \frac{[MS]}{[M][S]} \qquad [3.16]$$

(5) Solid phase

$$MA_n <===> M^{+n} + nA^{-}; \qquad K_{sp} = \left[M^{+n}\right]\left[A^{-}\right]^n \qquad [3.17]$$

where:

K_j is the concentration equilibrium constant for the metal-hydroxyl complex equilibrium; j is a numerical coefficient.

K_i is the concentration equilibrium constant for the metal-ligand complex equilibrium; i is a numerical coefficient.

K_A is the concentration equilibrium constant for the adsorbed metal equilibrium; A stands for adsorbed metal.

Ksp is the concentration equilibrium constant for the solid phase equilibrium; sp stands for solid phase.

See figure 3.12 for a conceptual model of metal speciation and distribution.

Figure 3.12
Resonance Structures for O_3

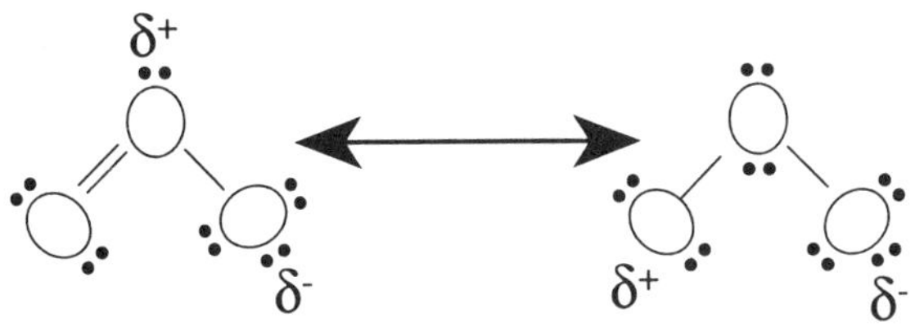

From Bablon, G., Bellamy, W.D., Bourbigot, M., Daniel, F.B., Dore, M., Erb, F., Gordon, G., Langlais, B., Laplanche, A., Legube, B., Martin, G., Masschelein, W.J., Pacey, G., Reckhow, D.A., Ventresque, C., *Fundamental Aspects*, 11, 17, 21, in *Ozone in Water Treatment: Application and Engineering*, Langlais, B., Reckhow, D.A., Brink, D.R., Eds., Lewis Publishers, a subsidiary of CRC Press, Boca Raton, Florida, 1991. With Permission.

3.5 **Status Of Development**

Except for the PEG Process, the substitution processes described herein are available for commercial application. As discussed earlier, a pilot system of the PEG Process was tested at the site in Guam but no further activity occurred. Table 3.7 (on page 3.62) lists the treatment projects conducted by the GRC Process. As can be seen, most of the treatments have been performed on either PCB-contaminated liquids or at pilot-scale on contaminated soils. Table 3.8 (on page 3.63) shows the processes conducted using high temperature substitution.

According to the developer, the low temperature substitution reactions have not been highly successful in laboratory tests for the treatment of halogenated aliphatic compounds. The high temperature processes, because they rely on pyrolysis as well as on substitution reactions, should be effec-

Table 3.7
Sites of Operation of GRC Process

Site	Date	Contaminant	Size/Concentration
ITS	1992	PCB Soil	150-ton demo.
Wide Beach	1988	PCB Soil	10 runs @ 200 lb each 260 to< 2 ppm
AmTech	1988	dioxin Sludge	2500 gal @1.1% to <1ppb
GE, Moreau	1987	PCB Soil	Pilot Test 7000 to <10 ppm
NCBC	1987	Dioxin Soil	Pilot Test 350 ppb
Montana Pole	1986	Dioxin Oil	9000 gal 100 ppm to <1 ppb
Western Processing	1986	Dioxin Oil	5500 gal @ 15% water, to <1 ppb
Niagara Mohawk	1986	PCB Oil	6000 gal to <2 ppm
Bengart-Memel	1986	PCB Soils	12 tons to <10 ppm
Niagara Mohawk	1984-5	PCB Oil	Pilot Test to <2 ppm

Other demonstrations conducted under EPA sponsorship.

tive in treating aliphatics. Since the SoilTech/ATP process uses the chemistry of the BCD Process to treat the chlorinated organic compounds, similar results are expected for both.

3.6 Environmental Impact

The impact of chemical treatment processes on the land and water is limited to that of the material leaving the process, either treated soil or treated water. Posttreatment of the soil or aqueous streams can limit the impacts to acceptable levels. The impact of oxidation or precipitation processes on the air is slight to nonexistent, making them highly attractive because air pollution control devices can increase the cost of competing technologies by a factor of two or more.

3.6.1 Substitution Processes

The potential effect of substitution processes on the air must be carefully evaluated. The processes can and should be designed to essentially eliminate air emissions, because they will not do it naturally. It is generally assumed that contaminants, such as PCBs, are nonvolatile at temperatures at which low-temperature destruction processes operate (130 to 160°C (270 to 320°F) for PCBs). This assumption is wrong; PCBs do have finite vapor pressures at these temperatures.

A review of the Alternative Treatment Technology Information Center (ATTIC) data base (US EPA 1992a) revealed that the vapor pressures of common Aroclors (commercial mixtures of PCBs and chlorobenzene isomers) at 100°C are in the range of 0.5 to 3 Torr (mm Hg). Mid-weight Aroclors, those that were most commonly used, have a vapor pressure of about 1 Torr. Assuming ideal vapor pressure behavior, whenever a mixture of water and a PCB are heated to just 100°C, 1 mole of the PCB will be evaporated for every 760 moles of water. This ratio is independent of concentration of PCBs. Perry (1965) provides for more information on steam stripping and steam distillation. A typical molecular weight for an Aroclor is approximately 270. The molecular weight of water is 18. Hence, at 100°C, for every kg of water evaporated, 1/760 x 270/18 = 0.02 kg or 20 g of PCBs will also evaporate.

Table 3.8

Sites of Operation, High Temperature Chemical Treatment Processes

Site	Date	Process/ Contaminant	Size/ Concentration
Wide Beach Development Superfund Site, Brant, NY	1991-'92	ATP/PCB	40,000 yd^3, Soil 100-600ppm
Waugegan Harbor, IL	1992	ATP/PCB	Demonstration 10 tph dredge spoils 99.9999% destruction
Australia	1993 (sch.)	BCD/PCB	2-2,000 L batch systems for liquids
U.S. Navy	1993 (sch.)	BCD/PCB	1-tph soil 25 - 6,500ppm

sch. - scheduled for the projected date

This calculation is based on the assumption that the water does not change the PCBs' vapor pressure. In fact, PCBs are highly hydrophobic, and the presence of water will result in a vapor pressure that is higher than this value. Thus, the volatilization rate of PCBs is, most likely, higher than the above estimate. In addition, it is expected that the vapor pressure of PCBs will increase at a greater rate than that of water as the temperature increases. In conclusion, significant quantities of PCBs will volatilize in any process operating at a temperature above 100°C. High temperature substitution processes take advantage of this phenomenon, collecting the condensate, separating the oily phase which contains the PCBs, and recycling the collected oil back into the reactor.

3.6.2 Oxidation Processes

The typical compounds identified after natural waters are ozonated are saturated aliphatic acids and diacids, aliphatic aldehydes, alkanes, and some aromatic acids. These compounds are also identified in oxidation by $KMnO_4$ or in hydrolysis by NaOH. These results suggest that the ozonation of complex solutions may not result only in oxidation products, but may cause the release of compounds from macromolecular structures or generate by-products that are susceptible to subsequent hydrolysis or precipitation.

While ozonation destroys the mutagenic and carcinogenic activity of many compounds (e.g., polyaromatic amines and polycyclic aromatic hydrocarbons), it has also been observed to change the route of a compound's mutagenicity and produce mutagenic by-products. Two by-products shown by the Ames test to have direct mutagenic effects are glyoxal and glyoxylic acid. Overall, however, the ozonation of water is considered to produce far fewer potentially deleterious by-products than chlorination.

Ozone in the gas phase has long been recognized as a dangerous and toxic compound. In animal studies, inhalation of molecular ozone imparts chromosomal damage and inhibits DNA replication. In mammals, ozone's high redox potential and extreme reactivity result in damage to many biochemical components. Since the lung is vulnerable to gaseous ozone, ozonation must be conducted in closed systems and the offgas must be destroyed. The effects of aqueous ozone on mammals have not been extensively studied. Studies of several species of fish have shown ozone to be lethal to a number of them, but there is not much toxicity associated with common by-products.

Ozone is a stronger oxidant than the halogen/halide redox couples and thus, it is thermodynamically possible to oxidize chloride, bromide, and iodide. In the presence of bromide ions, for instance, ozonation leads to the formation of hypobromite, which is partially oxidized into bromate ions. In the presence of NOM, ozonation will produce bromonated organic compounds, such as, bromoform, bromoacetic acid, and dibromoacetic acid. The generation of these compounds presents the same set of concerns as other halogenated organic compounds because of their potential toxicity.

One advantage of direct UV photolysis is that complete destruction of compounds can be accomplished without any threatening chemical residue, if the energetic specifications for the photodissociation of the contaminant are met. Since this requires a special set of conditions (pulsed UV continuum or certain mixtures of compounds (dioxins + solvents)), more often parent compound degradation creates a suite of by-products. Careful attention should be paid to the identity of these compounds, which, in many cases (TNT, PCBs, dioxins), is extremely difficult to determine. Frequently, additional treatment, such as bioremediation or carbon adsorption, will be required, especially when solvents are present. If dangerous by-products accumulate, addition of oxidants to the UV system will usually result in more complete destruction of the parent compound and its breakdown products.

3.7 Pre- and Posttreatment Requirements

Liquid streams fed to oxidation processes are pretreated by filtration to remove sediment and other particulates. Oxidation processes using UV light are particularly sensitive to turbidity; if the light cannot penetrate the liquid, loss of efficiency will result. Deposits on the UV lamps will also result in a decrease of process efficiency. Some deposition and fouling is unavoidable, and a well-designed system will make provisions for regular cleaning of the lamp surfaces. In cases of extreme deposition, some form of pretreatment of the incoming aqueous stream may be necessary to control fouling.

No general guidance to posttreatment of the aqueous streams issuing from oxidation processes can be given. The requirements are highly site specific. Consider for example, an oxidation process used as a method of treating groundwater from a pump-and-treat operation. If the groundwater is contaminated with only small amounts of chloromethanes or chloroethanes, posttreatment will usually be unnecessary. If, on the other hand, the contaminants include highly refractory compounds, the oxidation process might be used to pretreat the compounds and make them more amenable to biological treatment.

3.7.1 Substitution Processes

Contaminated oil to be treated by substitution processes usually does not require pretreatment. Reagent usage, however, can often be reduced and process performance improved by dewatering and filtering the oil prior to treatment.

The key to a successful treatment of contaminated soil is to maximize the contaminant's accessibility to the reagent. Therefore, delumping and screening prior to treatment is highly desirable.

Posttreatment of soils or other waste streams is generally not required with the commercially available processes, as they incorporate the posttreatment in the operation itself. Needless to say, new processes that treat contaminated soils using substitution reactions (such as the BCD Process) must incorporate sufficient posttreatment of the soil to neutralize excess caustic and remove contaminants and reagents.

3.7.2 Oxidation Processes — Posttreatment Requirements

Since ozone is both a toxic gas and a fire hazard, it is mandatory that the residual O_3 in the effluent and offgases of an ozonation system be destroyed before release. There are three basic methods for destroying the ozone:

- Thermal treatment at temperatures between 300 to 350°C (570 to 660°F) for 3 seconds. This will destroy 99% of this ozone. Heat recovery may be incorporated in this type of system because of the resulting high temperatures and energy consumption;
- Catalytic thermal destruction wherein lower temperatures can be used (30 to 50°C (90 to 120°F)). Catalysts may be metal (platinum or palladium) or metal oxides (aluminum or manganese

oxide) for full-scale applications and hydroxides or peroxides for pilot scale. The main disadvantage of this method is catalyst sensitivity to chlorine, sulfides, and nitrogen; and

- Destruction of ozone by passing the gas stream through an activated carbon bed or column.

Posttreatment of by-products, of course, varies with their nature. For example, waters treated with ozone often require biostabilization because the biodegradability of the background organic material is enhanced. This can be effected through use of biological activated carbon or other types of biological filters. A thorough evaluation to determine by-product identities and toxicities should be made before releasing effluents into the environment.

In most systems where UV photolysis has been applied, posttreatment was not required. This is largely because such systems are used to treat relatively low concentration streams and, in a situation where incomplete destruction was observed, oxidants could probably be added to insure complete contaminant destruction via an alternate reaction pathway. It is possible, however, that for concentrated or complex systems, posttreatment of the effluent, by biological methods for example, would be desirable. Posttreatment may be necessary for applications where polymerization occurs. Polymerization of the organics would form a precipitate that would necessitate a solid-liquid separation step.

It is postulated that the UV/H_2O_2 treatment of organic contaminants in water produces a variety of small molecule organic acid intermediates which ultimately can be mineralized upon extended oxidation (Ogata, Tomizawa, and Takagi 1981). This must be verified, however, for the particular matrix, contaminant, and reactor system. As a precaution, offgases and effluent streams from the UV/H_2O_2 oxidation process should be discharged through a secondary carbon bed filtration system to account for process failures or effluent variability. The final effluent composition should be monitored to ensure that discharge requirements are met. When designed and operated properly, the UV/H_2O_2 oxidation process should not have any adverse environmental impact. The effluent from the UV/H_2O_2 oxidation process may require pH adjustment, solids removal, and the removal of residual hydrogen peroxide.

If engineered so that proper dose is applied, AOPs are capable of effecting very high levels of contaminant destruction with virtually no by-products. The only major posttreatment step required is the destruction of ozone in the offgas. This step is usually a part of the commercial process, and, in the Ultrox systems, (see Subsection 3.2.1) the gas treatment system is also capable of treating any VOCs that carry over in the offgas.

3.8 Special Health and Safety Considerations

Special worker health and safety considerations consist of those normally required in handling (1) PCBs and other materials of environmental and health concern and (2) corrosive materials, both caustic (the reagent) and acid (for neutralization) (American Chemical Society 1979; US Dept. of Health, Education, and Welfare 1977).

The long-term effects of the products of the substitution reactions are not certain. Specifically, the health effects of substituted PCBs, chlorodibenzofurans, and chlorodibenzodioxins are not known. All processes, except the KPEG Process used in Guam, either destroy the products in the system (BCD, ATP) or collect the product for subsequent proper disposal (GRC, KGME/DECHLOR).

Oxidative processes use strong oxidizing agents to break down the organic constituents, introducing a significant health and safety concern. The resulting materials handling requirements are reasonably well understood. The processes also use highly alkaline materials to treat chlorinated organics. Sodium and potassium hydroxide can react with aluminum, forming hydrogen. Any alkaline process should be designed to safely vent or control hydrogen gas. Even though use of aluminum in the treatment system must be scrupulously avoided, aluminum metal is often present in contaminated soil and debris.

The potential dangers in using ozone are even greater when oxygen is used as the feed gas. Although the system can be shut down if a leak develops, other safety measures should be taken, such as installation of self-contained breathing apparatuses.

3.9 Design Data and Unit Sizing

3.9.1 Substitution Processes

The parameters under which the substitution processes described herein operate do not impose special design considerations. For example, the residence times of one to three hours in the reactor for the GRC, KGME/DECHLOR, ATP Reactor, and BCD chemical processes are long enough to allow the reactions to occur unless some form of interfering material is present. Because of the possibility that some material may be present that could interfere with the chemical reactions, it is necessary to conduct treatability studies to evaluate the efficacy of the process. The high level of control that substitution reactions allow provides a high degree of certainty that the results of a properly-designed treatability study will translate reliably to field conditions.

The unit sizing or throughput of the process is largely determined by such factors as the type of soil or oil and the presence of water. Contaminant concentration is only a factor when it is very high.

Reactor size is determined largely by solid or slurry residence time. For all substitution processes, solids residence times are in the one to three-hour range. The reagent to soil ratios are from 1:1 to 2:1 by weight for the low-temperature KPEG Process, whereas, for the high-temperature BCD Process, the catalyst PEG weight ratio is 1 to 10% for a 50 lb process. For a 10 lb reaction in the KPEG Process, the ratio is 0.1 to 10%. The excess reagent is necessary as a phase transfer catalyst, which extracts the contaminant from the soil and allows it to react with the reagent. The reactor volume must be large enough to accommodate both the soil and reagent. The reactor for the low-temperature substitution processes must also be large enough to allow sufficient agitation

In essence, the substitution process "reaction" requires, along with the proper temperature, sufficient residence time in the reactor and adequate mixing in the vessel. It is also necessary to continuously collect the condensate, which may, or may not, be recycled in the reactor depending upon its makeup and value. After these steps, the treated effluent is discharged.

3.9.2 Oxidation Processes

The UV/hydrogen peroxide oxidation process is carried out in a recirculating batch reactor or continuous flow-through reactor configuration. System design is dependent upon the residence time requirements of the specific oxidation being undertaken and the level of degradation to be achieved. An analysis of the configurations for lamp arrays and reactor shapes in photochemical reactor design that optimize UV light flux has been reviewed by Smith (1988). Galvanized and aluminum reactors displayed no observable rate differences (Barr 1976). The internal reflectivity of the reactor vessel chamber may be an important design feature of the system, if significant quantities of light pass through the reaction medium.

The premixed hydrogen peroxide aqueous influent (with the pH adjusted, if needed) is pumped through the reaction chamber and passed by a series of UV lamps sealed in quartz tubes. Although the UV/H_2O_2 oxidation process is effective over a relatively broad pH range, the photolytic degradation of hydrogen peroxide is both mildly pH and temperature dependent. The rate of hydrogen peroxide photolytic decomposition is approximately twice as fast at pH 11 than at pH 3 (Sundstrom and Klei 1986) and increases by a factor of two in going from 17°C (63°F) to 35°C (95°F). The residence time in the reaction vessel must be sufficient to achieve complete oxidation of the initial contaminants and their oxidation by-products. Residence times from several minutes to several hours may be required.

Lamp geometry is another design consideration. Cylindrical lamps can be placed in the center of annular reactors or around the circumference of cylindrical reactors. Quartz must be used for those surfaces through which UV light passes before coming into contact with contaminants. On small scales, processing chambers can be placed alongside lamps and encircled by a UV reflector. This is the configuration used with a pulsing UV continuum in the Wekhof Process. The choice of geometry depends largely on the degree of light penetration (or attenuation) through the fluid. Since UV light cannot penetrate particles, a thin film flow over, for example, an irradiated weir could be used. For soils of shallow depths, frequent turning and the addition of either solvents or surfactants to bring contaminants to the surface are usually required.

The kind of lamp used determines the wavelength of the resultant UV light. It is necessary to match the wavelength to the reagents as closely as possible. Hydrogen peroxide absorbs light most efficiently at 254 nm.

More than 80% of the total output of low pressure mercury arc lamps is centered at a wavelength of 254 nm, making them preferable for most hydrogen peroxide photochemical processes.

Clearly, the greater the optical path length, the more likely a quantum of light is to strike the desired target molecule and initiate the chemical reaction. Filtration or other means of clarifying the influent is, therefore, a necessary part of any process utilizing UV light.

The aqueous stream to be treated is typically premixed with a 50% solution of hydrogen peroxide and fed into an oxidation reaction chamber containing the UV light source. Generally, excess hydrogen peroxide is used in the process, and the initial level of peroxide added is established in a laboratory- or bench-scale treatability study. The hydrogen peroxide level is directly proportional to the rate of oxidation observed and the level of peroxide added may require adjustment according to compositional changes in the influent or reactor performance.

The amount of hydrogen peroxide used depends upon the balance between reaction rate and peroxide efficiency. It is generally best to start the process at the lowest effective level of peroxide that results in complete mineralization and then, to increase its concentration to achieve the desired reaction rate. Excess hydrogen peroxide must be destroyed or removed from the effluent prior to discharge.

In general, AOPs operate at ambient, or slightly above ambient, temperatures and produce activated oxygen species, such as the hydroperoxy and hydroxyl radicals (Aieta et al. 1990). The term "activated oxygen species" is loosely applied to the following array of species: singlet oxygen, superoxide anion, peroxy radicals, hydroxyl radical, hydrogen peroxide, and organic hydroperoxides. It is important to note that a high degree of interconversion exists among these species, and the rates of production depend on the reaction conditions (Singh 1986).

Ozone is a powerful oxidizing agent, widely used in the treatment of water. Initially, because of this quality, its principal use was in disinfecting drinking water, but over time O_3 was found to be highly effective in many other water treatment applications. Ozone is now considered a multipurpose treatment chemical, and its use as an integrated process can effect removal of color, control of taste and odor, oxidize iron and manganese, enhance coagulation, control algal growth, minimize production of disinfection by-products, and provide biological stabilization.

The mechanisms by which ozone accomplishes these various treatment objectives are, for the most part, poorly understood. It is generally believed that ozone modifies organic structure in a way that enhances the overall treatment or removal process. The lack of knowledge results mostly because the chemical quality of natural waters, especially the nature of the natural organic matrix, is extremely difficult to define. Based on fundamental chemical considerations, however, ozone is known to be very reactive with a broad range of synthetic organic compounds (SOCs), and there is degree of certainty that ozone should be effective in destroying certain types of SOCs.

In aqueous solutions, there are two types of interactions between ozone and various compounds:

- Direct molecular reaction; and
- Indirect reaction with radical by-products of ozone decomposition.

In the first route, direct interaction (see figure 3.13) is highly selective. It occurs only at certain sites or functionalities and only under certain solution conditions. The second pathway, indirect interaction, is nonselective and is thought to be mediated primarily by the hydroxyl free radical.

Figure 3.13
Distribution of Heavy Metals Under Equilibrium Conditions

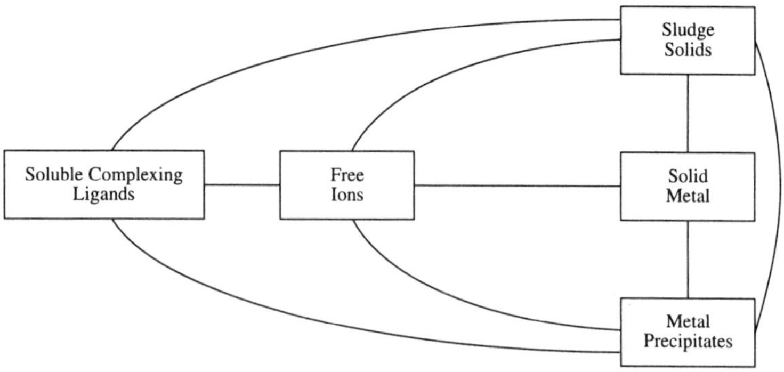

Figure 3.14
Schematic Representation of Direct and Indirect Reaction Pathways of O_3

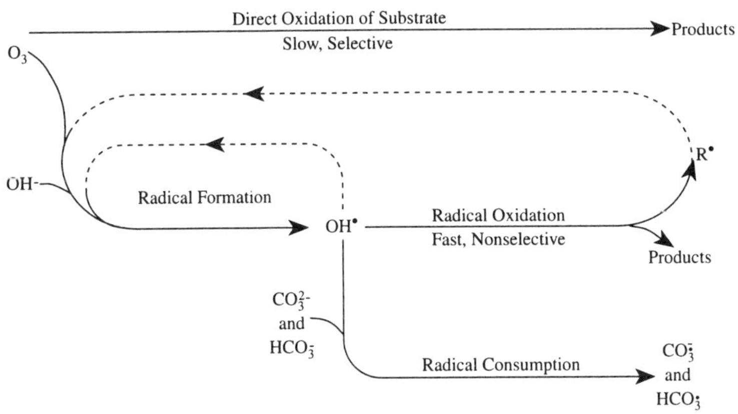

Reprinted from Water Research Volume 10, Number 5, J. Hoigne, and H. Bader, Role of Hydroxyl Radical Reaction in Ozonation Processes in Aqueous Solutions, Copyright 1976, with kind permission from Pergamon Press Ltd., Headington Hill Hall, Oxford, 0X3 0BW, UK.

Given the resonance structures of ozone, shown in figure 3.14, it is evident that ozone possesses the attributes of a dipole, an electrophile, and a nucleophile. Ozone's dipolar structure makes it capable of 1-3 dipolar cyclo addition to unsaturated bonds, also referred to as the Criegee mechanism, ultimately yielding carbonyl compounds (aldehydes and ketones) and hydrogen peroxide. Electrophilic reactions with ozone are limited to molecules having strong electronic densities, such as certain aromatic compounds. The substituents on aromatic rings exert a major influence on ozone attack. Aromatic substitution by electron donor groups such as OH or NH_2 causes an enhanced electronic density on carbons located in ortho and para positions, which, in a sense, activates these sites to electrophilic attack by O_3. Electron withdrawing substitutions such as COOH and NO_2 have an opposite effect on aromatic reactivity with ozone and direct attack to the least deactivated position that is located meta to the substitution.

Electrophilic interactions lead to hydroxylated products that will readily undergo further reaction with ozone. Typically, aliphatic products with carbonyl and carboxyl functional groups are among the final products. Nucleophilic interactions are common at sites showing electronic depletion, most often produced by electron withdrawing groups. Overall, interactions with molecular ozone are highly specific, slower than those of the indirect pathways, and favored by low pH (J.M. Montgomery Consulting Engineers 1985). The ionic state of a substrate has significant influence as well. Rates of interaction with molecular ozone are greater for non-ionic and anionic species than for cationic compounds. Rate constants for direct reaction with ozone in water have been measured for 45 potential organic contaminants (e.g., solvents, haloalkanes, esters, aromatics, and pesticides). These data illustrate that steric factors are also important in limiting reactivity for complex molecules (Yao and Haag 1991). In general, direct reaction with ozone is limited to unsaturated aromatic and aliphatic compounds and is influenced by the type and position of functional group substitutions (Langlais, Reckhow, and Brink 1991).

Under ambient conditions, ozone is relatively unstable and will rather quickly undergo decomposition. The stability of dissolved ozone is a function of pH, ultraviolet light, ozone concentration, and the concentration of radical scavengers. Ozone decomposition is faster under alkaline conditions and occurs in a chain reaction process. It is generally accepted that O_3 decomposition is base-catalyzed, and that the free-radical initiating step is the rate-determining step in the process. Although the hydroxide ion is considered the primary initiator under the process conditions in many water systems, there are many other compounds that can initiate the process, which involves inducing the formation of a superoxide ion radical, O_2:. In addition to hydroxide, some common inorganic ions that can serve as initiators are hydroperoxide and ferrous ions. Under acidic conditions, the O atom formed in the thermal dissociation of ozone is a precursor of the initiation of O_3 decomposition (Sejested et al. 1991). Examples of initiating organic compounds are glyoxylic acid, formic acid, and humic substances. Ultraviolet radiation at 253.7 nm also causes ozone decomposition, as does the combination of H_2O_2/HO_2-. These latter initiators are often used in combination with ozone and constitute the basis of AOPs (Langlais, Reckhow, and Brink 1991).

Promoters of free-radical reactions are all those inorganic and organic compounds capable of regenerating the superoxide ion from the hydroxyl radical. Common promoters are compounds such as aryl groups, formic acid, glyoxylic acid, primary alcohols, humic substances, and phosphates.

The superoxide ion will, in turn, react with ozone, playing a promoter's role.

Inhibitors of free-radical interactions terminate chain reactions by consuming hydroxyl radicals without regenerating the superoxide anion. Typical inhibitors are bicarbonate and carbonate ions, alkyl groups, tertiary alcohols, and humic substances. The observation that humic substances are capable of initiating, promoting, and terminating the chain reaction of O_3 decomposition results because these compounds are complex macromolecules substituted with many different functional groups.

Since the rate-limiting step in thc O_3 decomposition chain reaction is the initiation step, AOPs have been developed to enhance decomposition kinetics. These techniques are particularly effective at neutral pH and when high concentrations of radical scavengers (inhibitors) are present. The two most frequently used AOPs combine O_3 with either hydrogen peroxide or UV light.

As a weak acid, H_2O_2 partially dissociates in water into the hydroperoxide ion, HO_2^-. In contrast with the slow reaction between H_2O_2 and O_3, HO_2^- is highly reactive with O_3. The increasing O_3 decomposition rate that occurs with increasing pH is further accelerated in the presence of H_2O_2, since its equilibrium is shifted toward the conjugate base, HO_2^-. Figure 3.11 (on page 3.59) is a diagram of H_2O_2/HO_2^- induced decomposition of ozone.

The overall reaction in the formation of ozone is: $3\,O_2 \longleftrightarrow 2O_3$. This is a thermodynamically unfavorable reaction requiring a large input of energy. Unlike molecular oxygen, ozone cannot be liquefied by compression. It can be dissolved in liquid oxygen up to 30% by weight, but beyond that it becomes explosive. For these and other reasons, ozone is generated on site in most applications.

The production of ozone requires the dissociation of molecular oxygen into oxygen radicals, which then react with molecular oxygen to form ozone. Ozone generation is an equilibrium process in which conditions for generation also influence destruction. Oxygen radicals, for example, promote the destruction, as well as production, of ozone. Therefore, it is important to determine and maintain an optimum concentration of oxygen radicals in order to produce efficient conversion.

The splitting of oxygen requires significant inputs of energy. Electrical discharges or photon quantum energy are typical energy sources. The high voltage source of electrons most widely used for ozone generation is silent corona discharge, but chemonuclear sources and electrolytic processes are sometimes used. Ultraviolet light, of wavelengths lower than 200 nm, or gamma-rays are possible photon quantum energy sources.

In the corona discharge, a high voltage alternating current (6 to 20 kv) is passed across a dielectric discharge gap containing a dry, oxygen-bearing feed gas. The efficiency of ozone generation depends on the type of feed gas used. Ozone concentrations of 1.5 to 2.5% by weight can be achieved by using air, and by using high-purity oxygen, ozone concentrations can be increased to the 3 to 5% range. Pretreatment of air or high-purity oxygen feed gas removes dust, moisture, oil, and, in some cases, nitrogen. The elimination of moisture is imperative in order to obtain high yields. When air is used, water vapor promotes the formation of nitric acid, which causes corrosion, and production of hydroxyl free radicals, which consume oxygen radicals and ozone. Use of an air feed is complicated, costly, and maintenance-intensive. Use of liquefied oxygen eliminates most of the pretreatment needs and, at a cost of approximately $0.08/m^3$ ($3.00/ft^3), may be an economically attractive alternative, especially for small to mid-size applications (Kawamura 1991).

There are three basic kinds of ozone generators used in water treatment applications: low-, medium- and high-frequency units with variable or constant voltage. The low- and medium-frequency units are reliable, widely used, and readily supplied. Less heat is generated with these units, so cooling requirements are also decreased. Cooling systems are integral parts of ozone generator designs, since 90 to 95% of the supplied power is converted to heat. Power consumption is an important consideration because, in general, heat generation increases with power consumption.

Ozone is not highly soluble in water. Dissolution of ozone follows Henry's Law and, as such, is proportional to the partial pressure of ozone in the gas phase. Given ozone's low solubility and low concentration in the gas phase, its transfer from the gas to the liquid phase is a critical step affecting process efficiency. Following are the several kinds of contactors designed to optimize gas transfer:

- concurrent and countercurrent diffused bubbles;
- positive pressure injection (U-tube);
- negative pressure (Venturi tube);

- turbine mixer tank; and
- packed tower.

Tank depths are in the range of 5.5 to 6 m (18 to 20 ft), effecting a transfer efficiency of at least 95%. With high ozone concentrations, approximately 10% by weight, transfer can be accomplished with a hydraulic eductor and an in-line static mixer. The countercurrent bubble contactor has been used most often in water treatment because of its efficiency and cost-effectiveness. Ozone contact tanks are covered to contain the offgas and are normally built of concrete.

Extensive modelling has been done in evaluating gas transfer as a function of different operational parameters and reactor design (Langlais, Reckhow, and Brink 1991). Typically, however, since ozone contact produces a minimum of 95% gas transfer and the reaction kinetics of ozone are very rapid, contact times of 3 to 10 minutes are considered sufficient. Pilot testing is performed where time and budget permit, but selection of dose and contact time are often based on rule-of-thumb.

Since 3 to 10% of the ozone is not transferred to the liquid, ozone in the offgas may be present at a level of 1 g/m^3 and must be destroyed. In an eight-hour work day, the maximum ambient ozone concentration allowable by the Occupational Safety and Health Administration (OSHA) is 0.002 g/m^3. Techniques available for destroying the offgas are thermal destruction, with or without a catalyst, and catalytic thermal destruction. Simple thermal destruction is used in air-feed gas systems. Catalytic thermal destruction improves energy efficiency, but care must be taken to protect the catalyst from exposure to chlorine and its derivatives, sulfides, and nitrogen. In some cases, the offgas is recycled back to the feed, but this requires reconditioning.

In summary, the design of ozonation processes involves the selection of:

- a feed gas system;
- feed gas pretreatment;
- an ozone generator;
- contactor; and
- offgas destruction system.

An excellent discussion of design criteria is provided by Kawamura (1991).

Destruction efficiency of UV photolysis depends on the type of organic compound. The rate of the photodissociation process, Γ, can be calculated according to the following:

$$\Gamma = \Phi \sigma n[mg / Ls]$$

where: Φ is the photon flux within an absorption band (photons/cm^2·sec)

σ is the cross-sectional area for a photodissociation of a contaminant (10^{-17} to 10^{-16} cm^2) calculated for each absorption band.

n is the contaminant concentration (mg/L)

At a UV flux of 0.1 W/cm^2/nm (typical for the Wekhof Process) a rate of toxic destruction, Γ = 1 mg/L-sec, is possible. Although similar fluxes can be obtained with traditional mercury vapor lamps, the emission lines typically do not match the absorption bands of the various contaminants. Therefore, it is wavelength, not power, that is limiting.

One way of improving UV efficiency is by using pulsed UV sources. Pulsed sources have the advantage of producing pulses of very high photon fluxes that can be thousands of times greater than emissions from continuous sources. The pulsed sources operate at about the same average power level as the continuous sources, but they do so in short pulses of high intensity radiation. The short, high intensity pulses favorably alter the kinetics of the photochemistry. There are basically three design parameters for photon fluxes: peak power, RMS power, and average power. Peak power is defined as the energy of a single pulse divided by the duration of the pulse and is usually in orders of magnitude greater than the average power. The RMS power expresses the effectiveness of the repeated pulsed action of the peak power at some repetition rate, R. The average power is the combined energy of all the pulses delivered in one second at the specified repetition rate.

Optimum destruction of contaminants is accomplished by determining the proper combination of these parameters within the following ranges — the ratio of RMS power to average power should fall into a range of 1:10 to 1:100 and the ratio of average power to peak power should be within 1:1000 to 1:10,000 with an average power density of 0.1 W/cm^2/nm in the treated medium. The latter ratio is related to the plasma temperature and shifts the peak of UV generation to a absorption band region for a targeted contaminant (Wekhof 1991).

3.9.3 Precipitation Processes

The key to effective precipitation lies in assuring that the solubility products of the target species and the reagent used to precipitate them are below their concentrations in the solution. See table 3.9 (on page 3.80) for a list of values of solubility products of various metal hydroxide, metal carbonate, and metal sulfide species.

When a solid phase is precipitated from solution, impurities that are normally soluble under the conditions of the precipitation may adsorb onto nuclei or crystals and be removed with the parent solid as a single phase. This phenomenon is known as coprecipitation and is a concern when designing a precipitation system involving multiple substances.

Coprecipitation/adsorption is a coprocess for removing contaminants from wastewaters. The following five major kinds of coprecipitation have been identified (Kolthoff 1932; Salutsky 1959; Christian 1977; Patterson 1988):

- *Surface adsorption.* Impurities are not incorporated into the internal crystal structure, but instead stay adsorbed to the outer surface of the precipitate. This adsorption involves a primary adsorbed ion layer that is held tightly, and a counterion layer that is held more or less loosely. Surface properties of the forming solid phase (including electrostatic charge) serve to attract or repel secondary constituents in the surrounding aqueous matrix;
- *Occlusion.* Impurities are not incorporated in the crystal lattice, but are adsorbed during the growth of the crystals and give rise to the formation of imperfections in the crystal. Adsorption phenomena during the growth of the crystals are primarily responsible for the degree of occlusion;
- *Isomorphic inclusion (or mixed crystal formation).* The impurity fits nicely into the crystal lattice of the precipitate and becomes incorporated into the lattice in place of a lattice ion of similar dimension and chemical characteristics. Thus, the impurity becomes permanently incorporated into the crystal lattice, resulting in a mixed crystal;

Table 3.9
List of Solubility Products for Various Heavy Metal Compounds

Compound	pK_{sp}	K_{sp}
$CdCO_3$	11.28	$5.2x10^{-12}$
$Cd(OH)_2$ - fresh	13.6	$2.5x10^{-14}$
$Cd(OH)_2$ - aged	13.7*	$2.0x10^{-14}$
CdS	26.1	$8.0x10^{-27}$
$CoCO_3$	12.84	$1.4x10^{-13}$
$Co(OH)_2$ - fresh	14.8	$1.6x10^{-15}$
$Co(OH)_3$	43.8	$1.6x10^{-44}$
a-CoS	20.4	$4.0x10^{-21}$
B-CoS	24.7	$2.0x10^{-25}$
$Cr(OH)_3$	30.2	$6.3x10^{-31}$
	30.17*	$6.7x10^{-31}$
$CuCO_3$	9.86	$1.4x10^{-10}$
	9.60*	$2.5x10^{-10}$
$Cu(OH)_2$	19.66	$2.2x10^{-20}$
	18.80*	$1.6x10^{-19}$
CuS	35.2	$6.3x10^{-36}$
	36.10*	$8.0x10^{-37}$
$FeCO_3$	10.50	$3.2x10^{-11}$
	10.70*	$2.0x10^{-11}$
$Fe(OH)_2$	15.1	$8.0x10^{-16}$
	14.74*	$1.8x10^{-15}$
$Fe(OH)_3$	37.4	$4.0x10^{-38}$
	37.22*	$6.0x10^{-38}$
FeS	17.2	$6.3x10^{-18}$
	18.39*	$4.0x10^{-19}$
Hg_2CO_3	16.05	$8.9x10^{-17}$
$Hg(OH)_2$	23.7	$2.0x10^{-24}$
Hg_2S	47.0	$1.0x10^{-47}$
HgS (red)	52.4	$4.0x10^{-53}$
HgS (black)	51.8	$1.6x10^{-52}$
$MnCO_3$	10.74	$1.8x10^{-11}$
	9.40*	$4.0x10^{-10}$
$Mn(OH)_2$	12.72	$1.9x10^{-13}$
	12.70*	$2.0x10^{-13}$
MnS - amorphous	9.6	$2.5x10^{-10}$
MnS - crystalline	12.6	$2.5x10^{-13}$
MnS	15.15*	$7.0x10^{-16}$
$NiCO_3$	8.18	$6.6x10^{-9}$
	6.85*	$1.4x10^{-7}$
$Ni(OH)_2$ - fresh	14.7	$2.0x10^{-15}$
	15.80*	$1.6x10^{-16}$
a-NiS	18.5	$3.2x10^{-19}$
B-NiS	24.0	$1.0x10^{-24}$
V-NiS	25.7	$2.0x10^{-26}$
NiS	20.52*	$3.0x10^{-21}$
$PbCO_3$	13.13	$7.4x10^{-14}$
	12.82*	$1.5x10^{-13}$
$Pb(OH)_2$	14.93	$1.2x10^{-15}$
	14.38*	$4.2x10^{-15}$
PbS	27.9	$8.0x10^{-28}$
	28.15*	$7.0x10^{-29}$
$Sn(OH)_2$	27.85	$1.4x10^{-28}$
$Sn(OH)_4$	56.0	$1.0x10^{-56}$
SnS	25.0	$1.0x10^{-25}$
$SrCO_3$	9.96	$1.1x10^{-10}$
$ZnCO_3$	10.84	$1.4x10^{-11}$
	10.52*	$3.0x10^{-11}$
$Zn(OH)_2$	16.92	$1.2x10^{-17}$
	16.35*	$4.5x10^{-17}$
a-ZnS	23.8	$1.6x10^{-24}$
B-ZnS	21.6	$2.5x10^{-22}$
ZnS	22.8*	$1.6x10^{-23}$

Adapted from Dean 1979, and Benefield, Judkins, and Weand 1982
*Benefield, Judkins, and Weand 1982.

- *Mechanical entrapment.* This form involves the physical enclosure of a small portion of the mother liquor with tiny hollows or flaws which form during the rapid growth and coalescence of the crystals. The pockets remain filled with the mother liquor and eventually become completely enclosed by the precipitate; and
- *Postprecipitation.* The precipitate is allowed to stand in contact with the mother liquor, and a second substance will slowly form a precipitate with the precipitating reagent. This type of precipitate contamination is closely related to surface adsorption.

Regardless of the kind of coprecipitation, the initial incorporation of the impurity into the solid phase is the result of adsorption. This adsorption may be due to chemisorption, resulting from the coordination between the impurity and one or more constituent ions of the crystal lattice, or physisorption, resulting from electrostatic interactions, Van der Waal's forces, or dipole-dipole interactions. Chang (1985) and Chang and Peters (1985) presented data on the coprecipitation/adsorption of cadmium, lead, and zinc onto $CaCO_3$ sludges.

Patterson (1988) observed that little effort had been made to control coprecipitation. Possible control mechanisms include pH control (which influences the surface charge of the precipitate solid and the speciation of the soluble phases), control of the oxidation state of the soluble species, selection of coprecipitant salt and dosage, and process configuration.

3.9.4 Materials of Construction

In designing chemical processes, materials of construction often present a challenge, since the materials may be exposed to the following:

- highly reactive reagents;
- solvents specifically chosen to dissolve heavy organic compounds;
- agents specifically designed to mobilize metals; and
- highly abrasive conditions.

Highly reactive reagents are of particular concern when selecting elastomeric seals and gaskets for chemical treatment processes for halogenated organic compounds. The reagents are chosen to specifically attack the organic halogen. Most common elastomers, specifically designed for chemically resistant seals and gaskets, are based on halogenated polymers. Many dechlorination reagents will, for example, attack fluorochlorocarbon (i.e., Teflon™) gaskets, even though the material is recommended for use in contact with some of the solvents used in such a reaction.

Metals used in construction of oxidative processes must be selected for their resistance to oxidation. The first choice, therefore, is some type of stainless steel; however, many stainless steels are especially prone to chloride embrittlement. In addition, many stainless steels are susceptible to acid damage. Therefore, if the pH cannot be maintained in an acceptable range, a material that resists acid attack must be used.

While low pH is most commonly associated with corrosion problems, high pH is also a matter of concern. Chemical dechlorination (i.e., sodium adduct, KPEG, etc.) involves the use of highly alkaline reagents which will aggressively attack aluminum and magnesium. Many nonferrous metals are attacked by strong caustic agents; that most commonly encountered is aluminum, which is often used in tankers because of its light weight. Aluminum tankers can, inadvertently, end up as part of the treatment train in a site remediation. Copper is attacked by chloride salts, as well. The designer must be aware of these possibilities and guard against them.

Finally, the designer must consider mechanical wear as well as chemical resistance. Many wastes contain high concentrations of suspended solids. Many of the solids, especially in remediation wastes, are of a mineral nature and are highly abrasive. This can create a particularly acute problem with seals. Soil treatment systems are especially susceptible to abrasive attack. Furthermore, seal failure is especially dangerous in a chemical treatment system, since the system may contain highly reactive materials.

3.10 Operational Requirements and Considerations

As does any on-site process, chemical treatment processes require utilities and cooling water at the site. Chemical processes are typically compact and can be relatively self-contained. Thus, they do not require extensive site preparation or construction.

3.11 Materials Handling

The chemical treatment process selected must satisfy engineering requirements that are usually dictated by materials handling considerations. This is especially true for solids treatment, but materials handling can be a significant consideration in treating aqueous streams as well. The mixtures of waste and reagent can be abrasive and corrosive. Some processes may require that treatment take place at pressures other than atmospheric, and, therefore, the waste and reagents must be fed through pressure locks. Each process dictates unique materials handling techniques. In addition, the process must accommodate wastes of varying characteristics.

Another important consideration in materials handling is the control of fugitive emissions. Conveyors and other materials moving equipment from the construction or minerals-processing industries must often be modified for use in waste treatment systems so that fugitive emissions are minimized. A conveyor that releases a small amount of dust may be acceptable for transporting gravel, but its potential use in transporting a contaminated material must be carefully evaluated. Fugitive emissions present a particular problem for chemical treatment systems. They do not incorporate a means of treating air from hoods or shrouds, because unlike incinerators, they usually do not require large volumes of air. A covered conveyor transferring waste to an incinerator can be ducted to the incinerator's air intake duct. Use of the same conveyor in a chemical treatment system might dictate that the air from it be ducted to an air pollution control device.

Contaminated wastes and materials are, by their nature, highly heterogeneous. Any process designed to cope with an "average" material, no matter how well the waste is characterized, must also be capable of treating the off-average material to some extent. If not, frequent breakdowns will result. Preprocessing (e.g., blending and shredding) can help, but rarely can it provide a truly homogeneous material continuously over time.

Inhomogeneity of wastes usually translates into an unacceptable level of downtime. To avoid this problem, often the wastes to be treated are sampled over space and time. Each sample is analyzed or samples are combined and analyzed. The results of the analyses are then used to establish the "waste characteristic," and the treatment process is built to handle this average material.

Inhomogeneity of field materials presents problems for all treatment processes, but the problem is especially acute in chemical treatment processes, since the chemistry is often highly sensitive to trace materials in the waste. In selecting a process, it is essential that one examine the wastes to be treated in order to determine difficulties likely to be incurred under alternative treatment schemes.

Materials handling does not present a particular problem for the oxidation and precipitation processes addressed herein, except, possibly, in the collection and handling of precipitates from these processes. (Note that precipitates can form in oxidation as well as in precipitation processes.) The handling of precipitates present widely varying problems, and no general guidance can be given. Their handling must be part of the treatability study conducted before the treatment method is selected.

3.12 *Information Required to Consider and Employ the Process*

Following are factors that must be considered in deciding whether any given site or waste is a candidate for chemical treatment by a substitution process:

- accessibility of contaminants to the treatment process;
- materials handling;
- materials of construction;
- coupling with other processes;
- heterogeneity of inputs;
- treatment objectives; and
- cost-effectiveness.

The treatment objectives must be clearly established before a process is chosen. The level of destruction of the target contaminants is important, but the treatment objectives must be extended to other contaminants, both hazardous and nonhazardous. This is important when considering any remediation process, but especially when considering chemical treatment processes, since in most of them the contaminated material is mixed with reagents.

The kind of issues surrounding treatment objectives can be illustrated by considering treatment using KPEG to destroy chloro-p-dioxins (dioxins) on soil. The contaminated soil is mixed with the KPEG reagent, diluted in a solvent, and the mixture is heated to approximately 200°C (390°F). Once the dioxins are reacted, the mixture is cooled. It is now a slurry of KPEG, solvent, soil, and the products of the chemical reaction.

The mixture is highly alkaline and must be neutralized with acid. This step is clearly necessary because the alkaline material is hazardous and cannot be released. The neutralization step converts the excess KPEG to a potassium salt and the original polyethylene glycol, PEG. The PEG is not hazardous; in fact, PEGs of lower molecular weight are used as food additives. The solvent, however, may be hazardous. The question that now must be addressed is, how much subsequent treatment is required to remove the PEG and solvent so that the treated soil may be disposed?

The PEG is costly, so a high degree of PEG recovery is essential for the process's economics. One process recovers the vast majority of the PEG and solvent with a second treatment step, soil washing. But this step does not remove all of the PEG or solvent. Does the soil from the soil washing process require additional treatment (say, by bioremediation) prior to disposal? These questions illustrate how the selection of a given treatment scheme, especially one involving chemical treatment, hinges on the level of desired performance. It also illustrates that the designer must look beyond the environmental impact of just the one or a few contaminants that are of particular concern in a given situation. In many situations, the concern lies with a relative few toxic contaminants, but the designer must consider also the nontoxic pollutants such as biochemical oxygen demand (BOD), chemical oxygen demand (COD), NO_x, particulates, etc.

By its nature, chemical treatment requires that reagents be brought into intimate contact with the contaminant. To illustrate, consider PCBs on soil. Soil is a complex, highly variable combination of silicates, carbonates, organic matter, and numerous other constituents. In addition to chemical differences among soils, the physical states also vary widely. The particles can be coarse or fine, and they can be solid, highly porous or anything in between. Naturally-occurring soils will usually contain significant quantities of water. A typical clay or loamy soil that appears dry to the touch can contain in excess of 20% water by weight.

Contaminated soils are even more complex. As an example, consider a soil contaminated with a hydrophobic material, say PCBs. The following points should be taken into account when possible cleanup procedures are considered:

- The PCBs contaminate the soils in two unique ways: (1) on the surface; and (2) embedded in the pores of the soil particles;
- Only rarely are PCBs found in the environment in pure form. Most PCBs were blends of PCB isomers and chlorobenzenes. The blends are usually mixed with mineral oils that were used as dielectric fluids as well. The mineral oils were either paraffinic (aliphatic) or aromatic based;
- Generally, PCBs in the environment have "aged," changed with time. The aging consists of some oxidation of the less stable organic constituents, especially the mineral oils, dechlorination, and biodegradation of the lower chlorinated PCBs, and preferential volatilization of the lighter components. The combination of these processes leads to a change in the PCB isomer mix remaining in the soil; and
- The PCBs and the mineral oil base in which they are typically found are highly hydrophobic. Typical maximum solubility in water is on the order of 0.05 to 0.1 mg/L.

It is reasonable, therefore, to postulate that a grain of soil in a typical environmental sample of soil contaminated with a PCB will resemble that in figure 3.15 (on page 3.87). The soil grain's surface may be coated with the PCB liquid. The PCB liquid is also present in its pores, but it appears likely that the pores also contain water, creating a pattern of alternating layers of PCB and water.

Now, assume that a hypothetical reagent that readily destroys PCBs has been discovered. Fundamental kinetics dictate that the reagent and the PCB must be in the same phase. If this does not occur, then the chemical reaction (no matter how fast its rate) will be heterogeneous. The reaction rates will be limited by mass transfer of the PCB across the two-phase barrier and the destruction will be very slow.

If, on the other hand, the reagent solution can dissolve the PCB, the surface of the PCB will be dissolved and will react. But if the solution is hydrophobic, it will be unable to penetrate the droplets of water in the soil

Figure 3.15
Cross-Section of a Grain of PCB-Contaminated Soil

pores. The PCB in the pores will not be accessible to the reagent and, therefore, it will not be destroyed.

Chemical processes that treat PCB-contaminated materials overcome the problem because of two characteristics: (1) they are soluble in both organic and aqueous phases and (2) they operate at temperatures above the boiling point of water so that the water and PCBs are driven out of the pores, bringing the PCBs into contact with the reagent. For example, APEG reagents dissolve in both water and the organic phases. As a result, they can diffuse into the pores and destroy the PCBs. Furthermore, the processes are run at temperatures above the boiling point of water.

Consider how these facts relate to the treatment of soils using solid reagents. First, consider the contaminant on the surface of the soil. The solid reagent reacts with the contaminant on the surface of the soil, but there must be a mechanism for the contaminant to transfer from the soil particle to the reagent particle. Such a transfer is not impossible, but it is very slow. Contaminant in the pores is simply not accessible to the reagent. In order for a solid reagent to work, the contaminant must be mobilized either by heat (which vaporizes it) or by a solvent. The solvent, termed a phase transfer catalyst, is, therefore, an essential ingredient for any liquid-phase solid-solid reaction.

Although the problem of accessibility is most obvious in soils, it occurs in all media. For example, oxidation of organic contaminants in aqueous media by ozone requires that the ozone cross the liquid-gas barrier. The rate of ozone diffusion is much slower than the oxidation reaction. As a result, the destruction of organics by ozone will generally be mass-transfer limited. Improvements in the system's performance will come by increasing the surface area of contact, either by decreasing gas bubble size or increasing the system pressure.

3.13 *Unique Planning and Management Needs*

Chemical treatment is effective in treating selected contaminants that are in dilute form in another medium, such as soil, water, or nonhazardous oil. Understanding the principal advantages, and disadvantages, of chemical treatment is the key to determining the conditions under which it should be applied. Following are the principal advantages:

- Relatively low capital cost. The treatment system can often be assembled from standard, off-the-shelf components;
- Very low levels of air emissions. Because most chemical treatment systems operate in the liquid phase, the potential for air emissions is minimized;
- Potential for a high level of quality control (QC). The reaction products can be tested and analyzed as part of an on-line QC program to assure satisfactory destruction of the contaminants;
- Relatively high degree of public acceptance. The processes do not suffer from a popular stigma as does, for example, incineration, largely because of the low levels of air emissions that result and the high level of quality control that can be exacted; and
- Comparatively small size. Typically, systems with significant throughputs can be mounted on trailers and operated at the contaminated site with minimal construction and little on-site assembly.

Following are the principal disadvantages of chemical treatment:

- Relatively high operating costs, especially for treating materials with high concentrations of contaminants. The cost of chemical treatment is highly sensitive to the amount of reagent consumed. Because highly skilled operators are required on site, labor costs also tend to be higher than for simpler technologies;
- Highly sensitive to the overall composition of the material being treated. Water or other materials present in the contaminated material may interfere with some of the desired chemical reactions. The stream(s) to be treated must be relatively homogeneous;
- Risk of undesirable side reactions. The long-term environmental and health effects of the products of some chemical reactions used in treating contaminants are not well known; and
- Residual reagents. Except for some of the processes used to treat aqueous streams with H_2O_2 or O_3, residual reagents remain in the treated stream. This is of special concern when treating soils where the reagents must be removed by treating them thermally, through solvent extraction, or through soil washing.

In view of its principal advantages and disadvantages, chemical treatment should be considered as a method of treatment at sites where one or more of the following conditions exist:

- The object is mainly the treatment of one or a few specific kinds of contaminants that can be chemically modified;
- Transport of the contaminants to an off-site location for treatment is precluded by the volume of material to be treated or because of other considerations;
- Established on-site processes, such as incineration, are not technically feasible or are otherwise precluded; and
- The quantity of material to be treated is small or the contaminant concentrations are low and the economics of treatment, therefore, favor the low capital, high operating cost approach.

Political and social factors, rather than economic considerations, are often determinants in the selection of a treatment process. The lowest cost, environmentally-sound remediation may not be acceptable for a given site.

In such circumstances, chemical treatment may be selected in place of other techniques, although the economics may not be in its favor.

The overriding planning and management need for effective implementation of any remediation process is proper scheduling. This need is especially acute, however, in planning chemical treatment, because the chemicals required are generally not locally available. Limited on-site storage space makes delivery schedules especially crucial to effective operations.

3.14 Cost

This section focuses on how the incremental cost of the chemical process affects total treatment cost. It is keyed to the major components of fixed and variable cost listed below:

- FIXED COST COMPONENTS
  - equipment amortization
  - sales
  - obtaining required permits and approvals
  - preparing the system for the specific job
  - shipping, setup, and knockdown
  - site capping and restoration
- VARIABLE COST COMPONENTS
  - reagent purchase
  - handling and disposal cost of treatment and treated materials
  - utilities
  - labor, travel, and subsistence
  - chemical analyses

3.14.1 Fixed Costs

Because of the unique nature of chemical treatment, many of these components weigh in differently than they would in other remediation processes, such as incineration or stabilization.

The first factor that impacts chemical treatment is the amortization of the equipment. As previously discussed, chemical treatment systems tend to be highly site specific. That is, a given treatment train incorporating chemical treatment methods will be assembled to work on a specific site and then dismantled. In most cases, it is unlikely that the same treatment train will be used at a second site. As a result, the treatment system must be amortized over only one job. The cost per ton of material treated by a system which costs, for example, $3 million to build and to treat 30,000 tons of material at a site will be $111/tonne ($100/ton). The cost per ton does not include the cost of money (interest on capital). If the remediation takes three years between the time the equipment is built and the remediation is completed, and we assume a 10% annual cost of capital, the amortized cost of the equipment for this remediation is $145/tonne ($130/ton) of material treated.

Since typical disposal costs are currently on the order of $111 to $550/tonne ($100 to $500/ton), the amortization cost for this hypothetical system makes a significant impact on the cost of remediation. Clearly, chemical treatment systems must either be usable at many sites, or they must be relatively inexpensive to build. At present, with the exception of the KPEG Processes, the former alternative does not appear likely. Chemical treatment trains are highly specialized. Therefore, in order for chemical treatment systems to be economically competitive, the equipment must consist of easily-assembled, readily-available, reusable components.

Cost of sales is far more difficult to quantify a priori than amortization cost and is not unique to chemical treatment methods. It is significant, however, and must be included as a part of the overall cost of treatment when planning a remediation project.

Also difficult to quantify, permit costs are usually a significant fraction of the overall treatment cost. Permit costs have two components, direct and indirect. The direct permit costs are preparation of the permit application and the required testing and reporting. The indirect components are the cost of delays and responding to public comment. Examination of the history of several chemical treatment applications (i.e. Weitzman 1982; Peterson 1986; and operators of the ATP/SoilTech process) has led to the conclusion that chemical treatment systems encounter far less public opposition and fewer delays than do incineration processes, resulting in lower permitting costs than incineration.

Chemical treatment systems tend to be compact. They typically do not require large foundations (as do transportable incinerators) and are built on standard over-the-road equipment. They are, therefore, relatively inexpensive to ship, setup, and knockdown. The total cost of these activities is typically on the order of \$10,000 to \$20,000. These estimates are based on the authors' field experience and discussions with vendors.

Site capping and restoration can be a costly part of a remediation, these costs are not unique to chemical treatment. They are highly dependent on the characteristics of the particular site.

3.14.2 Variable Costs

Reagent cost is unique to chemical treatment processes. It is usually a very significant factor in the overall cost. Consider the treatment of 27,000 tonne (30,000 ton) of soil. The KPEG Processes require that the soil be mixed with relatively large volumes of reagent dissolved in a carrier. If, for purposes of order-of-magnitude cost estimation, it is assumed that each ton of soil is mixed with one ton of reagent, then it will be necessary to process 27,000 tonne (30,000 ton) of reagent. Reagent recovery will be considered below. Typical reagent costs are \$2.25 to \$11.25/kg (\$1 to \$5/lb). Assuming the lower cost, and neglecting, for the moment, reagent recovery, the reagent cost for the treatment is \$60,000,000 or \$2,200/tonne (\$2,000/ton). Clearly, some form of reagent recovery is economically necessary. Ignore, for the purpose of this analysis, the environmental impact of such an enormous reagent release. At 90% reagent recovery, the cost of reagent is \$220/tonne (\$200/ton) of material treated; at 95% recovery it is \$110/tonne (\$100/ton), and at 99% recovery \$22/tonne (\$20/ton).

It is extremely difficult to maintain 99% reagent recovery for a highly heterogeneous process. Ninety-five percent reagent recovery is still very difficult to achieve, although more realistic than the 99% claimed by one vendor. One can refine the reagent cost estimate by making the analysis site- and technology-specific, but it is apparent that in all cases, the cost of reagent is a significant part of the total remediation cost.

The above analysis is based on the use of quantities of reagent that would normally be needed to treat contaminated soils. The relatively large reagent volumes are necessary to overcome the mass transfer problems. In water treatment, far smaller quantities of reagent are required, on the order of 0.1%, making the cost of reagent far more manageable; however, it is still a significant factor when the cost of chemical treatment is compared to the cost of alternative technologies for many applications.

For most chemical treatment processes, utility costs are relatively low, typically less than $1.10/tonne ($1.00/ton) of material treated.

Labor, travel, and subsistence costs for chemical processes are generally higher than for other remediation operations. This is because the typical chemical treatment process requires highly trained workers. It is usually impossible to hire them locally. One must, therefore, add approximately $150 per day for travel and subsistence, corresponding to $18.75 per hour of operation. Assume that a typical soil treatment system requires three operators per shift. Assume further an average cost per operator is $45/hr, fully loaded, and a throughput of 9 tonne/hr (10 ton/hr). Adding (and rounding) the travel and subsistence cost increases the cost of labor for the operation to $64/hr per operator, or a cost of labor of about $21/tonne ($19/ton), if the system is operating eight hours per shift. System malfunctions, delays, and breakdowns reduce the time for treatment and proportionately increase the labor cost. For example, a system which, because of breakdowns and other reliability problems, is operating only 50% of the time would have double the labor costs per ton than that based on an estimate of 100% reliability. If we assume 50% time on-line, the labor cost for chemical treatment would increase to $42/tonne ($38/ton).

The above cost estimate does not include the cost of labor for excavation and materials transfer, which would be required regardless of the technology used. This cost varies with process reliability and is another factor making process reliability an important consideration in the selection of a remediation process. (Treatment system reliability, of course, is an important factor in the evaluation of all remediation processes, not just chemical treatment.) The excavation and materials handling equipment and personnel stand idle, while the system is inoperative. The cost attributed to such a decrease in productivity should be factored into the total cost of the treatment process.

Chemical treatment requires frequent sampling of both feed and processed materials and analysis of the samples for key constituents. The sampling and analysis is necessary both to assure that the process is destroying the target contaminant and to monitor the general performance. These costs cannot be quantified here, but one must take them into account in assessing the total cost of a chemical treatment operation.

While system reliability has its greatest impact on labor costs, it also affects other costs, such as rental of tank trucks, compressors, generators, and other equipment. In comparative analyses of treatment methods, system reliability is often determinative. The more complicated the process, the more significant downtime is likely. Chemical treatment systems use relatively simple equipment. If the equipment is properly selected for the application, it is likely that the system will have less downtime than systems applying alternative innovative technologies.

3.14.3 Estimated Costs of Various Treatment Methods

Reliable cost data on treatment methods are very difficult to obtain, since the ultimate cost is highly site specific. See table 3.10 (on page 3.95) for available cost data.

3.14.4 Equipment Sizing and Cost - Oxidative Processes

Excellent discussions of the practical aspects of ozonation are provided by Kawamura (1991) and Langlais, Reckhow, and Brink (1991). The most practical ozone generators are either low- or medium-frequency. Selection of a generator is usually based on the following factors:

- reliability and maintenance;
- energy cost differential;
- turn-down ratio;
- cooling water temperature;
- benefits of oxygen-enriched feed gas; and
- owner preference.

Following are key design elements and operating factors affecting equipment sizing and cost:

- size of the ozone generators;

Table 3.10
Cost of Treatment

Process	Treatment Cost per ton of material treated	Source
PEG Process	unknown	
GRC Process	\$200-500	Vendor claim
DECHLOR/KGME	N/A	
ATP	\$254	Vendor claim
BCD	N/A	
Romulus Remediation	\$1,358	Field Data

N/A - Not available

- number of generators, cost of energy, and cost of O_2;
- type of feed gas treatment;
- reliability of each component;
- operation and maintenance costs;
- ozone contactor design;
- offgas treatment; and
- whether UV light or hydrogen peroxide is used to catalyze decomposition of O_3.

Major manufacturers of ozone generators are Quantum (Emery), Welsback Ozone System, Griffin, Technics, PCI Ozone and Control, Asea Brown Boveri (ABB), Infilco Degremont Inc. (IDI), Trailigas, Schmidding-Werke (Megos), Mitsubishi, and Toshiba. Design criteria and a detailed example of design calculations can be found in Kawamura (1991). An extensive review of the economics of ozonation systems is presented by Bellamy et al. (1991). See also figures 3.16 a and b (on pages 3.96 and 3.97).

Wekhof (1991) presents cost comparisons for a 100-fold destruction of 20 mg TCE/L and 20 mg benzene/L in a wastewater using the following

Figure 3.16a
Construction Costs of Ozonation Systems

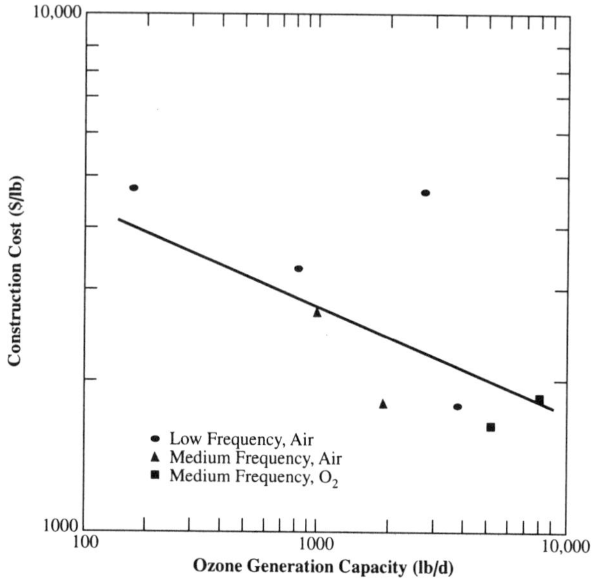

Complete ozone system (includes generation, building, and contactor).

Reprinted by Permission of Susumu Kawamura from "Water Treatment Principles and Design" by James M. Montgomery Consulting Engineers, Inc. Published by John Wiley and Sons. Copyright 1989 by Susumu Kawamura.

technologies: air stripping with offgas vapor phase activated carbon (Westates Carbon, Inc., Los Angeles, CA); conventional UV/hydrogen peroxide (PSI, Tucson, AZ); Wekhof Process using a 20 kW UVERG (pulsed UV lamp) system. The comparison is summarized in table 3.11 (on page 3.98). The pulsed UV system was considerably less expensive than conventional air stripping/offgas scrubbing. The system also enjoys an economic advantage over conventional advanced oxidation processes. For soils, the pulsed system's operational costs are much greater — in the range of \$120 to \$250/m^3 (\$92 to \$191/yd^3). It requires 24 hours to process 1 m^3 (1.3 yd^3) of soil using a 100 kW pulsed system and mobile systems are available. In contrast, treatment of highly contaminated soils by bioremediation takes between months and years and incineration is costly and destroys the soil.

The issue of equipment sizing is, as always, resolved through a site-specific tradeoff between investment capital and operating costs. Pilot- and commercial-scale UV/H_2O_2 oxidation systems for the treatment of contaminated groundwater have been designed and operated. Reactor capacities in the range of 57 to 830 L (15 to 220 gal) have been constructed and operated (Heeks, Smith, and Perry 1991; Andrews 1980). Flow rates through the oxidation reactors are designed to meet effluent contaminant levels by controlling residence times. UV/H_2O_2 reactor systems have been operated at flow rates of 4 to 37 L/min (1 to 10 gal/min). The selection of reactor volume, throughput, and residence time will be dictated by the rate of oxidation of contaminants, which can be controlled to a reasonable degree by the concentration of hydrogen peroxide in the influent. Operating systems have been constructed by several vendors including Peroxidation Systems, Inc. and Solarchem Environmental Systems.

Figure 3.16b

Breakdown of Ozonation System Construction Costs

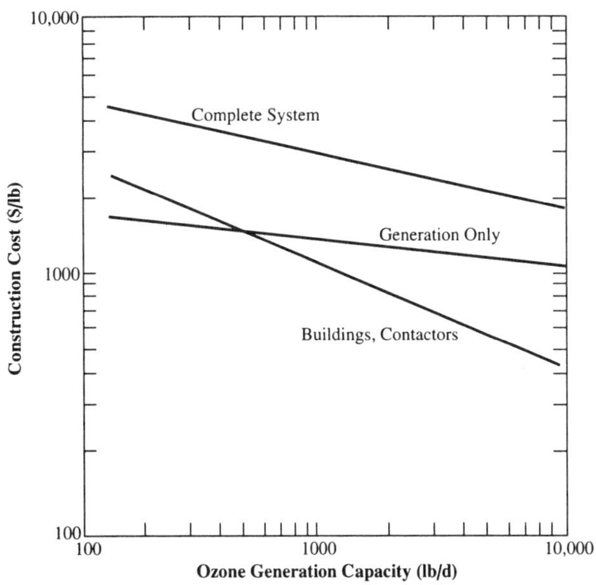

Ozone system components.

Capital and operating costs are dependent on the scale of the operation, contaminant levels, oxidation rates, post- and pretreatment and equipment utilization rates. Cost estimates for groundwater treatment of \$0.41 and \$1.32/m^3 (\$1.57 and \$5.00/1,000 gal) of wastewater treated at process throughputs of 1 m^3/min (250 gal/min) and 0.1 m^3/min (25 gal/min) for the treatment of groundwater containing approximately 6,000 ppm of VOCs have been generated (Heeks, Smith, and Perry 1991). A pilot-plant operation with a 0.04 m^3/min (10 gal/min) throughput rate has been projected to entail a capital investment of approximately \$100,000 and an annual operating cost of \$20,000.

Treatment costs, of course, depend largely on the specific set of conditions attending each application, such as, nature of initial concentration and contaminant, and flow and target concentrations. For initial concentrations less than 100 mg/L, the Rayox system reduces contaminant levels three to

Table 3.11

Economic Comparisons of Groundwater Treatment Systems

Item	Air Stripping And Activated Carbon	Mercury Vapor Lamp UV With H_2O_2	Pulsed Xenon Lamp UV With H_2O_2
Equipment	\$ 100,000	\$ 115,000	\$ 115,000
Installation	\$ 20,000	\$ 20,000	\$ 15,000
Capital Cost	\$120,000	\$ 135,000	\$ 120,000
Electricity	\$ 5,120	\$ 34,500	\$ 11,400
Carbon	\$ 274,300	\$ 0	\$ 0
Chemicals	\$ 0	\$ 12,000	\$ 6,000
Lamps Cost	\$ 0	\$ 38,400	\$ 17,800
Maintenance	\$ 15,600	\$ 6,750	\$ 6,000
Amortization	\$ 24,000	\$ 27,000	\$ 25,000
Total Operating	\$ 319,020	\$ 118,650	\$ 66,200
Total Cost			
\$/m^3	\$ 1.60	\$ 0.59	\$ 0.33
(\$/1,000 gal)	(\$ 6.07)	(\$ 2.24)	\$ 1.26

Costs projected according to the following assumptions:
Contaminant: 20 mg/L benzene plus 20 mg/L TCE
Flow rate: 380 L/min (100 gal/min)
Operating costs are given for 1 year round the clock operation

Reprinted by permission of the American Institute of Chemical Engineers from "Treatment of Contaminated Water, Air and soil with UV Flashlamps" by A. Wekhof. In Environmental Progress, Volume 10, number 4.

five orders of magnitude at a cost of approximately \$0.79/m^3 (\$3.00/1,000 gal). As initial concentrations increase from 100 to 1,000 mg/L, the costs rise to \$2.64/m^3 (\$10/1,000 gal).

Direct operating and maintenance costs of the Ultrox Process are reported by the manufacturer for a number of applications. The costs range from \$0.11 to \$1.32/m^3 (\$0.40 to \$5.00/1,000 gal). In general, it was found that this process was much less costly than alternatives, such as activated carbon. In one application, a comparison of various treatment strategies for groundwater contamination with TCE and other trace VOCs (total VOC concentration = 7.0 mg/L) found that the capital costs of air stripping with vapor phase granular activated carbon were less than the Ultrox Process, but the operation and maintenance costs of the Ultrox system were one-third to one-half that of the alternatives.

4 POTENTIAL APPLICATIONS

Chemical treatment processes are potentially applicable under the following conditions:

- Substitution processes — where materials are considered hazardous because they contain a specific class of contaminants, especially halogenated aromatic compounds. They also may be used in treating materials containing halogenated aliphatics, nitrogen-bearing compounds, and sulfonated compounds;
- Oxidation processes — where materials are considered hazardous because they contain low concentrations of organic constituents in water or in a dilute slurry; and
- Precipitation processes — where materials are considered hazardous because they contain toxic metal compounds in aqueous solution.

There are rules-of-thumb as to the degree to which some organic compounds are amenable to chemical treatment. Nonchlorinated organic compounds are commonly treated by oxidation reactions, while chlorinated compounds are treated by dechlorination/substitution reactions. See table 4.1 (on page 4.2) for lists of compounds in the approximate order to which they are amenable to treatment. Compounds falling into classes at the top of the table tend to be more readily treatable than those at the bottom. The nature of the contaminant to be treated is a major determinant whether chemical treatment is practical under any given condition.

As to inorganic contaminants, the question becomes whether a more desirable form of the metal to be treated exists chemically, i.e., different oxidation state, less soluble, more soluble, less toxic, etc. The determination must be made on a metal-specific basis and no general rules appear to apply. Organic compounds containing fluorine tend to be more stable and, hence, more difficult to treat chemically than their chlorinated counterparts.

Table 4.1
Amenable to Treatment (Rule of Thumb)

	Nonchlorinated (Oxidation Processes)	Chlorinated (Substitution Processes)	Media
Easiest to Treat	Phenols, Alcohols	Chlorophenols	Homogeneous Debris
↓	Ketones, Organic Acids		↓
	Aromatics	Chlor. Aromatics	
	Unsaturated Aliphatics	Chlor. Unsaturated Aliphatics	Sand
	Saturated Aliphatics	Chlor. Saturated Aliphatics	↓
Hardest to Treat	Flourinated Compounds	Brominated and Iodine Compounds	Clay

Brominated and iodinated compounds tend to be more amenable to chemical treatment.

The foregoing are merely guidelines based on the fundamental chemistry of the classes of compounds, supported, however, by the results of studies of the several treatment processes. To understand the likelihood of the success of a given chemical process, it is necessary to consider its scheme of attacking the target molecule. Consider, for example, a nonchlorinated phenol. It consists of a partially oxidized benzene ring. The -OH group on the benzene ring reduces the inherent stability of the benzene ring and makes the molecule more susceptible to further oxidation. Aromatic compounds that do not include a phenol group, such as benzene or naphthalene, are very stable and far more difficult to oxidize. The degree of difficulty progresses down the list.

Chlorinated (or, more generally, halogenated) organic compounds are typically treated by processes that attack the chlorine atom on the molecule, rather than oxidize the whole molecule. A possible exception is the lower chlorinated chlorophenols that can be oxidized by processes analogous to those for phenol. Clearly, chemical processes that attack the chlorine atom on the molecule are totally unsuitable for treatment of materials contaminated with nonhalogenated compounds.

There is also an important rule-of-thumb relating to the medium to be treated. Obviously, solids tend to be more difficult to treat than liquids.

Furthermore, the greater the contaminant concentration, the more difficult the contaminant is to treat. There are, however, treatability characteristics within the solids class. Table 4.1 (on page 4.2) lists various solid media in order of increasing difficulty of chemical processing; the larger the particle size, and the less hydrophilic the material, the easier it is to treat the media. This is because larger particles and less hydrophilic materials make the contaminant more accessible to the reagent and allow clean separation of the reagent from the matrix after treatment.

Within these classes of media, the difficulty of treatment increases with an increase in the water content of the solids. Contaminated sand as a dredge spoil would tend to be more difficult to treat than the same contaminated sand if it were removed from a dry area. The water tends to (1) dilute the reagents, (2) encapsulate hydrophobic contaminants, (3) make the solids clump together, and (4) interfere with many chemical reactions.

Although several oxidation processes have been applied directly to the treatment of contaminated soils on a laboratory scale, the oxidation reactions themselves are most likely to occur in the liquid phase and are more efficient in degrading aqueous phase organic compounds. Therefore, although direct treatment of contaminated soils on a large scale may prove to be physically impractical (Watts, Tyre, and Miller 1991), indirect treatment of soil contaminants with chemical oxidation may be feasible when phase transfer from the solid to aqueous phase can be effectively achieved. Chemical oxidation has also been evaluated as a pretreatment step in the biological degradation of contaminants in soil matrices (Kelly, Gauger, and Srivastava 1990; US EPA 1990).

Ozone is commonly used in water treatment to oxidize many inorganic compounds. Generally, the rate of oxidation is first-order with respect to both reactants. For many compounds, the rate varies with pH, as shown in figure 4.1 (on page 4.4).

Elevated levels of reduced iron and manganese are frequently found in groundwaters. While not posing a health threat, such concentrations are not desirable and interfere with other processes or uses of the groundwater. In a pump-and-treat scenario, for example, high iron and manganese levels in contaminated groundwater interfere with adsorption processes (e.g., activated carbon or ion exchange) and bioremediation techniques. Therefore, removal is required to assure the effectiveness of these processes.

Figure 4.1

Influence of pH on Rate Constants for Ozone and Various Inorganic Compounds

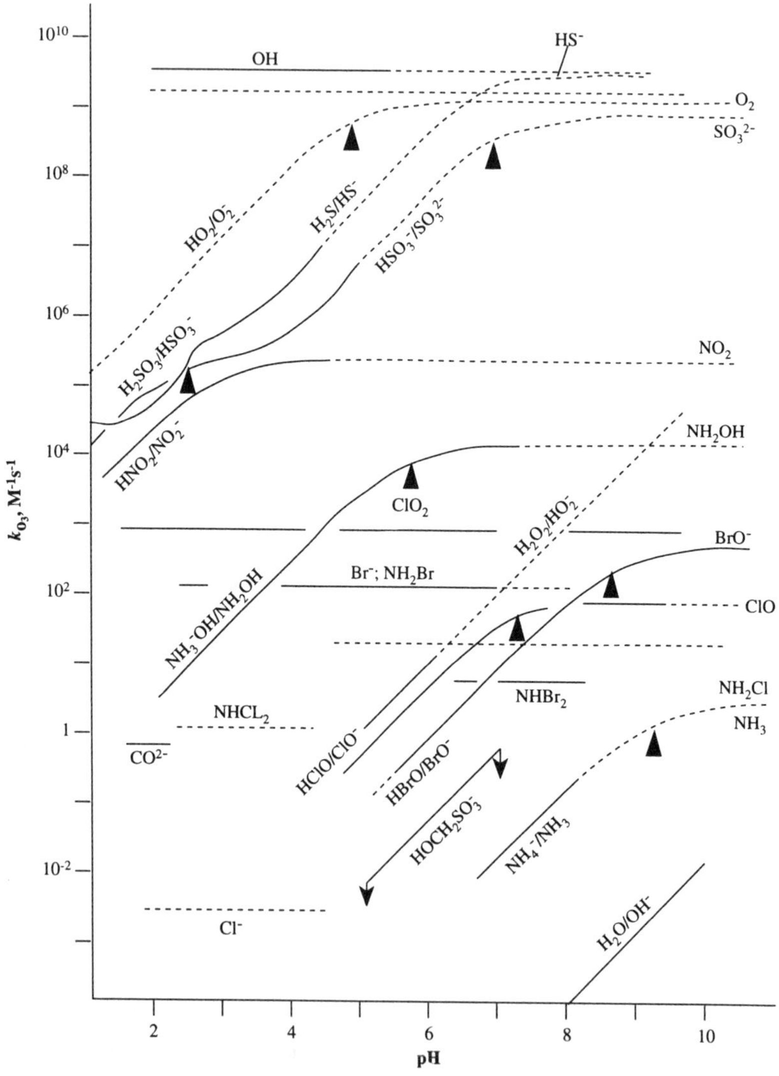

Reprinted from Water Research Volume 19, Number 8, J. Hoigne, H. Bader, W.R. Haag, and J. Staehelin, Rate Constants of Reactions with Ozone with Organic and Inorganic Compounds in Water, Part 3 Inorganic Compounds and Radicals, Copyright 1985, with kind permission from Pergamon Press Ltd., Headington Hill Hall, Oxford, 0X3 0BW, UK.

Reduced iron or the ferrous ion (Fe^{+2}) is easily oxidized by oxygen or ozone to Fe^{+3} in a stoichiometric ratio of 0.43 mg O_3/mg of Fe^{+2}. The oxidation of Mn^{+2} to Mn^{+4} requires stronger oxidation conditions than does the oxidation of iron, but ozone is sufficiently energetic. Actual removal of both iron and manganese is accomplished by the formation of an insoluble hydroxide precipitate. Therefore, in addition to ozonation, a solid-liquid separation unit is needed. There are many techniques to remove solids and, depending on flows and space availability, tray or tube settlers, a sedimentation tank, or a porous media filter may be appropriate. The oxidation and precipitation steps in this removal process will be influenced, or even severely hampered, by the presence of significant amounts of natural organic material. Higher ozone doses are required and, possibly, coagulant use would be necessary to destabilize the hydroxide precipitates.

The oxidation of ammonia to nitrate is possible with ozone added in a 4:1 molar ratio with ammonia.

$$4O_3 + NH_3 \rightarrow NO_3^- + 4O_2 + H_3O^+ \qquad [4.1]$$

The rate of this reaction is slow, particularly at pH values less than nine. Unless the system is adequately buffered, the pH decreases with reaction. Oxidation occurs more rapidly in the presence of bromide, which has a catalytic effect independent of pH. Nitrite, too, is quickly oxidized by ozone to nitrate and each mg of NO_2^- consumes 1.4 mg of O_3. Sulfide will be oxidized to sulfate at a rate of 6 mg O_3 per mg of S^{-2}. The reaction rate increases with the degree of deprotonation, or negative charge, of the inorganic species.

For the purposes of hazardous waste treatment, ozonation is probably more applicable to the destruction of synthetic organic compounds. Attention must be paid to the inorganic matrix because ozone demand will be exerted by the compounds described above. Carbonate species, the major ion of most water systems, are potent scavengers of hydroxyl free radicals. Both of these factors will diminish ozone's effectiveness and increase ozone consumption.

In natural waters, organic compounds are present in either a dissolved or particulate state. The concentration of dissolved organic carbon (DOC) in most natural surface and groundwaters is in the range of 2 to 10 mg/L and of this, 95% is derived from natural sources. Natural organic matter is composed of various biopolymers, such as amino acids and proteins, fulvic acids, carbohydrates, lipids, etc. Synthetic organic compounds are usually present in comparatively small proportions, around 5% of the DOC, and consequently are referred to as micropollutants. Micropollutants can be present as untransformed aromatic or aliphatic hydrocarbons, chlorinated solvents, phenols, substituted or nonsubstituted polyphenols, pesticides, plasticizers, or surfactants. When released into the environment, these compounds can be modified by hydrolysis, photolysis, or biological transformations to by-products that may be environmentally harmful.

Most synthetic organic compounds are hydrophobic and have very low solubilities in water. High concentrations of natural organic matter (NOM) can increase these solubilities either through partitioning, complexation, or by a cosolvent-like effect. Usually, synthetic organic compounds (SOCs) are found in higher concentration in the particulate organic phase (biomass), in sediments, and on soils. In complex media, meaning multicomponent and multiphase systems, reactivity with ozone must be very high for degradation to occur.

Direct reaction will occur at carbon multiple bonds (C=C, C=C-O-R, -C=C-X), or at atoms carrying negative charge (N, P, O, S, and nucleophilic carbons). Ortho- and para-activated aromatic compounds substituted with OH, CH_3 or OCH_3 are predicted to show strong initial reactivity in contrast to lesser reactivity, in cases where NO_2, COOH, or CHO are substituted. These chemical reactivity predictions would also hold for other electrophilic oxidants, but among these, ozone is more reactive. If the mode of reaction is due to radical species such as the hydroxyl radical, reactions are nonselective.

Continued reaction beyond the initial reaction depends upon the structure of the primary by-products, but, in most cases, additional reaction will occur, especially in the case of the radical pathway. The feasibility of using ozonation for organic destruction can be evaluated by considering relatively simple rules of organic chemistry. A very thorough review of ozone's behavior toward a wide range of organic compounds present in natural waters can be found in Langlais, Reckhow, and Brink 1991, the salient points of which are briefly summarized below.

Saturated aliphatic hydrocarbons and halogenated derivatives are largely an unreactive group of compounds with respect to direct interactions with ozone. Alkenes are more reactive, but this reactivity is depressed with substitution of electrophilic groups, such as halogens. Cleavage of carbon double bonds produces aldehydes and acids. In the case of ethylene, rates of reaction are much lower with hydrogen replacement by chlorine. In general, alcohols, ethers, aldehydes, and carboxylic acids are also relatively unreactive with ozone, and their degradation requires hydroxyl radicals. Destruction of these unreactive compounds requires advanced oxidation processes to stimulate radical reactions. Ozone, in combination with either H_2O_2 or ultraviolet (UV) light, would enhance the rates of destruction 1 to 50 fold.

Ozone reactivity with an aromatic compound depends on the compound's substituents. Interaction may occur with the ring, with side-chains, or competitively at both sites. To reiterate, direct ozone attack is most favorable as an electrophilic reaction at ortho or para positions relative to electron donor substitutions (e.g., OH), or at the meta position for electron withdrawing substituents (e.g., Cl^-). Mononuclear and polynuclear aromatic compounds are degradable with ozone. With naphthalenes, the first step in degradation is 1-3 cyclo addition to break one of the aromatic rings and subsequent reaction of the resulting side chains to produce a suite of by-products (Legube 1985, 1986). In general, the reaction scheme between ozone and mono- or polynuclear aromatic hydrocarbons involves oxidation of the aromatic ring to produce phenols, quinones, and/or aromatic acids. Further breakdown of the aromatic ring occurs, producing short-chain aliphatic acids and aldehydes. Less reactive aromatic compounds (chlorobenzenes, nitrobenzenes, chlorinated furans, and dioxins) require the use of advanced oxidative processes (AOPs) to direct the reaction efficiently along the free radical pathway.

Phenolic compounds are a common class of aromatic pollutants found in water. In general, phenolic compounds are susceptible to interactions with ozone due to the presence of the OH substitution, which is both electron-donating and ionizing. Therefore, reaction between ozone and phenolic compounds is favored by increasing pH; overall reaction rates have been seen to increase by an order of magnitude for each unit increase in pH. At higher pH, it is expected that the mechanism of O_3 interaction is a combination of direct and indirect reaction. Approximately 4 to 6 moles of ozone are required per mole of phenol for ring cleavage. While phenols are rapidly and easily oxidized by ozone, the suite of by-products can bear greater toxicity than the parent compounds. At sufficient ozone dose and contact times, these by-products are usually destroyed, yielding a final set of oxidation products including glyoxylic, oxalic and formic acids, and glyoxal. The extent and rate of reaction can be improved with AOPs.

Amines are another family of compounds that reacts with ozone. Electrophilic interactions are promoted by the electron-donating influence of nitrogen. This can be offset, though, by the presence of electron withdrawing groups such as nitro groups or nitrosamines. Analysis of by-products suggests that there are four types of interactions between amines and ozone:

- direct oxidation of the nitrogen group;
- oxidation of the α-carbon;
- deamination or splitting of the C-N bond; and
- secondary condensation or polymerization reactions among parent compounds and reaction by-products.

Rates of reaction with compounds such as urea and dimethyl nitrosamine are very slow and can be enhanced with advanced oxidation processes (see also Subsection 3.3.1, Photolysis Systems).

Ozonation of pesticides has been reasonably well studied, and a good review of the literature has been written by Reynolds (1989). The term *pesticide* encompasses a broad range of agents and chemistries. General predictions of the destructive capacity of ozone can be made by considering the organic structure of the pesticide. Organochlorine pesticides are not very reactive with ozone, and AOPs are usually applied to improve rates of destruction. Organophosphates, on the other hand, are much more reactive. The P=S bonds are oxidized to form P=O, and continued reaction yields phosphoric acid. Phenoxyacetic herbicides interact with ozone along lines similar to that of phenolic compounds (Benitez, Beltran-Heredia, and Gonzalez 1991). Heterocyclic nitrogenous herbicides, such as atrazine, are not very reactive. Direct interaction with pyridine is slow, but higher rates of reaction have been observed via the free radical mechanism which produces by-products that are very easily degraded (Andresozzi et al. 1991). A study, evaluating ozonation of a pesticide waste and rinsate (containing atrazine, cyanazine, metoalchlor, and paraquat) prior to circulation through a biologically active soil column, demonstrated that herbicidal activity was eliminated, and no mutagenic activity was found (Somich, Muldoon, and Dearney 1990).

Surfactants, as a group, are not particularly reactive with ozone. Their presence in water, even at relatively low levels (1 to 3 mg/L), interferes with gas transfer. There are four general classes of surfactants: nonionic, anionic, cationic, and ampholytic. A frequently used anionic surfactant that is fairly reactive with ozone is alkylbenzene sulfonate. In this case, ozone enhances the biodegradability of the compound. Reaction with anionic forms is favored over neutral and cationic surfactants. Increasing pH increases rates of reaction. In a study of the ozonation of polyethoxylated nonyl phenol, modification of the polyethoxylated side chain indicated that the reaction was essentially hydrogen abstraction at one of the ethoxylated units followed by depolymerization (Calvosa et al. 1991). Nonionic surfactants yield polyethylene glycols and polyethers as products of ozonation. These products, in turn, show very low reactivities with ozone. Overall, reactions between ozone and surfactants are slow, and in mixtures, other compounds of higher reactivity will consume ozone first.

Colored compounds such as dyes or NOMs typically display chemical attributes (anionic with conjugated aromatic structures), making them relatively reactive with ozone. Ozonation can result in some color loss, but can also yield a broad suite of by-products. The complexity of the molecules dictates dose and contact time requirements.

In summary, there are a number of trends that should be considered when assessing the usefulness of ozone in destroying SOCs:

- In engineering applications, significant degradation occurs only with compounds having ozonation rate constants, K_0, greater than 103 $M^{-1}S^{-1}$;
- Ozone reacts preferentially with nucleophilic sites (O, N, P, S substitutions, C=C);
- Destruction of parent compounds by ozone typically yields compounds more oxidized, polar, and biodegradable; the total organic carbon (TOC) of water, however, may not change. Therefore, monitoring TOC is not an adequate method of evaluating process efficiency;
- In some cases, ozonation by-products can have higher toxicities than the parent compounds. In addition, ozone can modify complexation behavior of trace materials (e.g. metals) causing their release into solution. Efforts should be made to evaluate by-product identity and toxicity; and

- Slow or poor reactivity with ozone can be enhanced by AOPs that involve combined use of O_3/H_2O_2, or O_3/UV, or O_3/H_2O_2/UV.

Ozone is a potent biocide and is often used as a primary disinfectant in treatment schemes. It is not considered an adequate final disinfectant and, because it is short-lived, it does not impart any residual disinfectant to water. Furthermore, ozone usually renders the natural organic matrix more biodegradable, which leads to biological regrowth in distribution systems. For this reason, biostabilization is often required after ozonation. While it is not certain how ozone inactivates microorganisms, it seems likely that the site of inactivation is the cytoplasmic membrane, and if residual ozone is able to cross this membrane, ozone would be highly reactive with the cytoplasm and nucleic acids.

Precipitation processes are applicable whenever the concentration of a target metal in an aqueous stream is greater than the solubility of its least soluble compound. Solubility products need to be considered when the stream contains more than one metal. See table 4.2 for a listing of sources reporting the successful removal of a variety of metals from aqueous streams by precipitation.

Table 4.2

Summary of Heavy Metal Removals Achieved Using Various Chemical Precipitation Techniques.

Metal	pH	Metal Concentration (mg/L) Initial	Residual	Removal Efficiency (%)	Comments	Reference
Hydroxide Precipitation						
Ba	10-11	7.0-8.5	—	>90	Pilot plant tests	EPA 1978
	9.2	10.0-12.0	—	84		
	10.5	10.0-12.0	—	93		
	11.6	10.0-12.0	—	82		
	10.5	7.5	—	88	Full-scale tests	EPA 1978
	10.3	17.4	—	95		
Cd	8.5-11.3	0.3	—	>98		
	11.2-11.3	10	—	>98		
Cd	6.0-10.0	100	<0.3	>99.7	Synthetic plating wastewater	Peters and Ku 1984
	10.0	100	<0.3	>99.7		
		100	2.0	98.0		
		100	5.0	95.0		

Table 4.2 cont.
Summary of Heavy Metal Removals Achieved Using Various Chemical Precipitation Techniques.

Metal	pH	Metal Concentration (mg/L) Initial	Residual	Removal Efficiency (%)	Comments	Reference
Cd	8.0	—	2350	—	10 mg/L CO_3^{-2}	Patterson et al. 1977
	8.6	—	126	—	10 mg/L CO_3^{-2}	
	9.4	—	5	—	10 mg/L CO_3^{-2}	
	10.4	—	0.2	—	55 mg/L CO_3^{-2}	
	11.9	—	0.3	—	100 mg/L CO_3^{-2}	
Cr	6.6	Cr^{+3}:5125	Cr^{+3}:26.0	>98.0	—	
	9.5-10.0	Cr^{+3}:1.40	Cr^{+3}:0.3	78.6	Nickel/Chrome rinse	Arumugam 1976
		Cr^{+6}:2.23	Cr^{+6}:<0.01	>99.5		
	9.5-10.5	Cr^{+3}:4.0	Cr^{+3}:0.03	99.3	Full scale plant	
		Cr^{+6}:4.5	Cr^{+6}:<0.01	>99.7		
Cr^{+3}	10.6-11.3	0.15	—	>98.0	—	EPA 1978
	9.5-10.5	0.15	—	>70.0	—	
Cr^{+6}	9.5-11.6	0.15	—	<10.0	—	
Cu	10.5	0.45	0.08	82.2	Copper rinse	Rabosky and Altares 1983
	9.5-10.5	5.7	0.89	84.4	Full scale plant	
Pb	8.5-11.3	0.15	—	>98.0	—	EPA 1978
	6.0	—	1700	—	15 mg/L CO_3^{-2}	Patterson et al. 1977
	7.4	—	25.6	—	15 mg/L CO_3^{-2}	
	8.8	—	6.0	—	15 mg/L CO_3^{-2}	
	10.5	—	0.6	—	40 mg/L CO_3^{-2}	
	11.9	—	280	—	123 mg/L CO_3^{-2}	
	12.3	—	1050	—	275 mg/L CO_3^{-2}	
Hg	10.7-11.4	9.3 µg/L	—	60-80	—	EPA 1978
	9.4	9.3 µg/L	—	30		
	9.3-11.3	9.3 µg/L	—	<5		
Ag	9.0	0.15	—	70	—	EPA 1978
	11.5	0.15	—	90		
Zn	10.0	100	0.3	99.7	<u>No chelants present:</u>	Peters & Ku 1985
		100	32.0	68.0	300 mg/L EDTA	
	6.2	—	1900	—	12 mg/L CO_3^{-2}	Patterson et al. 1977
	7.5	—	27.5	—	12 mg/L CO_3^{-2}	
	8.3	—	0.55	—	30 mg/L CO_3^{-2}	
	9.5	—	0.25	—	35 mg/L CO_3^{-2}	
	11.0	—	0.68	—	65 mg/L CO_3^{-2}	
	11.8	—	0.95	—	25 mg/L CO_3^{-2}	
					<u>No chelants present</u>:	
Zn	10.0	106	0.25	99.76	0.025-µm filter	Ku & Peters 1986
		106	0.34	99.68	0.45-µm filter	
		106	0.76	99.28	2.5-µm filter	
		106	46.0	56.60	11.0-µm filter	
					<u>100 mg/L Ammonia present</u>:	
	10.0	106	0.38	99.64	0.025-µm filter	
		106	0.40	99.62	0.45-µm filter	
		106	0.65	99.39	2.5-µm filter	
		106	41.0	61.32	11.0-µm filter	

Table 4.2 cont.
Summary of Heavy Metal Removals Achieved Using Various Chemical Precipitation Techniques.

Metal	pH	Metal Concentration (mg/L) Initial	Residual	Removal Efficiency (%)	Comments	Reference
					100 mg/L Citrate present:	
Zn	10.0	106	0.48	99.55	0.025-μm filter	Ku & Peters 1986
		106	0.88	99.17	0.45-μm filter	
		106	37.0	65.09	2.5-μm filter	
		106	>100	<5.66	11.0-μm filter	
					100 mg/L Tartrate present:	
	10.0	106	0.45	99.58	0.025-μm filter	
		106	1.04	99.02	0.45-μm filter	
		106	25.0	76.41	2.5-μm filter	
		106	>100	<5.66	11.0-μm filter	
					No chelants present:	
	10.0	106	>100	<5.66	5-min settling	
		106	5.8	94.53	15-min settling	
		106	4.5	95.75	30-min settling	
					100 mg/L Ammonia present:	
	10.0	106	>100	<5.66	5-min settling	
		106	7.0	93.40	15-min settling	
		106	6.2	94.15	30-min settling	
					100 mg/L Citrate present:	
	10.0	106	>100	<5.66	5-min settling	
		106	>100	<5.66	15-min settling	
		106	>100	<5.66	30-min settling	
					100 mg/L Tartrate present:	
	10.0	106	>100	<5.66	5-min settling	
		106	>100	<5.66	15-min settling	
		106	>100	<5.66	30-min settling	
Ni	6.8	—	1450	—	19 mg/L CO_3^{-2}	Patterson et al. 1977
	7.4	—	930	—	30 mg/L CO_3^{-2}	
	8.3	—	15	—	38 mg/L CO_3^{-2}	
	9.4	—	0.5	—	38 mg/L CO_3^{-2}	
	11.0	—	0.3	—	90 mg/L CO_3^{-2}	
	12.3	—	0.5	—	225 mg/L CO_3^{-2}	
Mixed Metals:	>8.5					
Cu		60.0	0.9	98.5	Electroplating plant data	Sheffield 1981
Ni		1.9	0.3	84.2		
Pb		1.2	0.4	66.7		
Mixed Metals:	8.5-91					
Fe		20-60	0.3-2.4	>96	Electroplating plant data	
Pb		0.1	0.5	—		
Sn		1.0	0.5	>50		
Mixed Metals:	8-10					
Cu		20-150	0.5-2.0	>97.5	Printed circuit board manufacturer	
Pb		0.5-1.0	<0.05	90-95		
Sn		0.1-0.5	0.1	<80		

Table 4.2 cont.
Summary of Heavy Metal Removals Achieved Using Various Chemical Precipitation Techniques.

Metal	pH	Metal Concentration (mg/L) Initial	Residual	Removal Efficiency (%)	Comments	Reference
Mixed Metals:	9.7-10.2					Brantner and
Cd		1.66	0.04	97.6	—	Cichon 1981
Cr		1.11	0.97	12.6		
Cu		0.29	0.03	89.7		
Pb		1.7	0.2	88.2		
Zn		31.0	0.28	99.1		
Carbonate Precipitation						
Cd	7.2	—	440	—	5 mg/L CO_3^{-2}	Patterson et al. 1977
	8.4	—	7.5	—	5 mg/L CO_3^{-2}	
	9.5	—	0.6	—	10 mg/L CO_3^{-2}	
	10.7	—	0.35	—	50 mg/L CO_3^{-2}	
	11.9	—	0.5	—	225 mg/L CO_3^{-2}	
Cd	8.1	—	5.0	—	1200 mg/L CO_3^{-2}	
	8.4	—	1.2	—	2800 mg/L CO_3^{-2}	
	8.7	—	1.7	—	3350 mg/L CO_3^{-2}	
	10.0	—	0.25	—	4200 mg/L CO_3^{-2}	
	10.8	—	0.25	—	4400 mg/L CO_3^{-2}	
	11.7	—	0.35	—	4300 mg/L CO_3^{-2}	
Ni	7.2	—	800	—	1500 mg/L CO_3^{-2}	
	8.2	—	60.0	—	3000 mg/L CO_3^{-2}	
	9.0	—	3.8	—	3500 mg/L CO_3^{-2}	
	10.5	—	2.7	—	5500 mg/L CO_3^{-2}	
	11.5	—	1.4	—	5500 mg/L CO_3^{-2}	
	12.5	—	1.4	—	5500 mg/L CO_3^{-2}	
Pb	5.6	—	4.6	—	10 mg/L CO_3^{-2}	
	6.1	—	43.6	—	15 mg/L CO_3^{-2}	
	6.8	—	17.4	—	20 mg/L CO_3^{-2}	
	10.1	—	0.6	—	55 mg/L CO_3^{-2}	
	11.6	—	10.0	—	55 mg/L CO_3^{-2}	
	12.3	—	1260	—	55 mg/L CO_3^{-2}	
Pb	7.5	—	1.0	—	1200 mg/L CO_3^{-2}	
	8.4	—	2.0	—	4000 mg/L CO_3^{-2}	
	9.2	—	3.6	—	4650 mg/L CO_3^{-2}	
	10.5	—	8.0	—	5500 mg/L CO_3^{-2}	
	11.4	—	6.0	—	5500 mg/L CO_3^{-2}	
	12.4	—	150	—	5500 mg/L CO_3^{-2}	
Zn	6.6	—	260	—	Sample not analyzed for CO_3^{-2}	
	8.3	—	0.95	—	Sample not analyzed for CO_3^{-2}	
	9.1	—	0.75	—	Sample not analyzed for CO_3^{-2}	
	10.0	—	0.60	—	5500 mg/L CO_3^{-2}	
	10.8	—	0.85	—	6500 mg/L CO_3^{-2}	
	11.9	—	1.60	—	5500 mg/L CO_3^{-2}	
	12.5	—	49.6	—	3500 mg/L CO_3^{-2}	

Table 4.2 cont.
Summary of Heavy Metal Removals Achieved Using Various Chemical Precipitation Techniques.

Metal	pH	Metal Concentration (mg/L) Initial	Residual	Removal Efficiency (%)	Comments	Reference
Mixed Metals: 7.8-8.5						Brantner & Cichon 1981
Cd		1.37	0.04	97.1		
Cr		0.67	0.60	10.4		
Cu		0.18	<0.03	>83.3		
Pb		1.4	<0.1	>92.9		
Zn		26	1.18	95.4		
Sulfide Precipitation						
Cu	4.0	100	0.08	99.92	No chelants, S^{-2} =1.05x	Peters et al. 1984a
	6.0	100	0.08	99.92	No chelants, S^{-2} =1.05x	
	8.0	100	0.05	99.95	No chelants, S^{-2} =1.05x	
	10.0	100	0.05	99.95	No chelants, S^{-2} =1.05x	
Cu	4.0	100	1.3	98.7	100 mg/L EDTA, S^{-2} =1.0x	Bhattacharyya et al. 1981b
	4.0	100	0.6	99.4	100 mg/L Gluconic Acid, S^{-2} =1.0x	
	4.0	100	0.3	99.7	100 mg/L Citrate, S^{-2}=1.0x	
	4.0	100	0.2	99.8	100 mg/L Tartrate, S^{-2}=1.0x	
Cu	3.0	100	0.9	99.1	100 mg/L EDTA, S^{-2} =1.05x	Peters et al. 1984a
	4.0	100	0.8	99.2	100 mg/L EDTA, S^{-2} =1.05x	
	6.0	100	0.6	99.4	100 mg/L EDTA, S^{-2} =1.05x	
	8.0	100	0.6	99.4	100 mg/L EDTA, S^{-2} =1.05x	
	10.0	100	1.0	99.0	100 mg/L EDTA, S^{-2} =1.05x	
Cu	4.0	100	0.08	99.92	No chelants, S^{-2} = 1.05x	Ku 1982
		100	0.85	99.15	100 mg/L EDTA, S^{-2} =1.05x	
		100	0.65	99.35	100 mg/L Citrate, S^{-2} =1.05x	
		100	0.25	99.75	100 mg/L Gluconic Acid, S^{-2} =1.05x	
		100	0.15	99.85	100 mg/L Tartrate, S^{-2} =1.05x	
	8.0	100	0.05	99.95	No chelants, S^{-2} = 1.05x	
		100	0.7	99.3	100 mg/L EDTA, S^{-2} =1.05x	
		100	0.4	99.6	100 mg/L Citrate, S^{-2} =1.05x	
		100	0.1	99.9	100 mg/L Tartrate, S^{-2} =1.05x	
Cu	8.0	100	0.5-1.0	>99.0	100 mg/L EDTA, S^{-2} =1.05x	Peters et al. 1984
Cu	4.0	100	0.08	99.92	No Chelants present S^{-2} = 1.05x	Peters & Ku 1987
	10.0	100	<0.05	>99.95	No Chelants present S^{-2} = 1.05x	
	4.0	100	0.65	99.35	100 mg/L Citrate S^{-2} = 1.05x	
	6.0	100	0.45	99.55	100 mg/L Citrate S^{-2} = 1.05x	
	8.0	100	0.40	99.60	100 mg/L Citrate S^{-2} = 1.05x	
	10.0	100	0.50	99.50	100 mg/L Citrate S^{-2} = 1.05x	

Table 4.2 cont.
Summary of Heavy Metal Removals Achieved Using Various Chemical Precipitation Techniques.

Metal	pH	Metal Concentration (mg/L) Initial	Residual	Removal Efficiency (%)	Comments	Reference
Cd	4.0-10.0	500	0.01	>99.99	S^{-2} = 1.05x	Peters et al. 1984a
	4.0	100	1.2	98.8	100 mg/L EDTA, S^{-2} = 1.0x	Bhattacharyya et al. 1981b
	9.0	100	0.16	99.84	No chelants, CaS precipitation	Kim 1981
Cd	4.0	100	0.1	99.9	No chelants, S^{-2} = 1.05x	Peters and Ku 1987
		100	0.5	99.5	100 mg/L Citrate S^{-2} = 1.05x	
	8.0	100	0.3	99.7	No chelants S^{-2} = 1.05x	
		100	0.9	99.1	100 mg/L Citrate S^{-2} = 1.05x	
Zn	3.0	100	12.0	88.0	No chelants, S^{-2} = 1.05x	Peters et al. 1984a
	4.0	100	0.3	99.7	No chelants, S^{-2} = 1.05x	
	8.0	100	0.2	99.8	No chelants, S^{-2} = 1.05x	
	10.0	100	0.15	99.85	No chelants, S^{-2} = 1.05x	
	4.0	100	16.5	83.5	100 mg/L EDTA, S^{-2} = 1.05x	
	6.0	100	15.0	85.0	100 mg/L EDTA, S^{-2} = 1.05x	
	8.0	100	12.8	87.2	100 mg/L EDTA, S^{-2} = 1.05x	
	10.0	100	12.0	88.0	100 mg/L EDTA, S^{-2} = 1.05x	
Zn	4.0	100	8.0	92.0	100 mg/L EDTA, S^{-2} = 1.0x	Bhattacharyya et al., 1981b
	8.0	100	12.0	88.0	100 mg/L EDTA, S^{-2} = 1.05x 120	Peters et al. 1984b
					No Chelants present:	
Zn	10.0	106	0.10	99.91	0.025-μm filter	Ku & Peters 1986
		106	0.10	99.91	0.45-μm filter	
		106	0.36	99.66	2.5-μm filter	
		106	26.0	75.47	11.0-μm filter	
					100 mg/L Ammonia present:	
	10.0	106	0.22	99.79	0.025-μm filter	
		106	0.24	99.77	0.45-μm filter	
		106	0.84	99.21	2.5-μm filter	
		106	25.0	76.42	11.0-μm filter	
					100 mg/L Citrate present:	
	10.0	106	0.35	99.67	0.025-μm filter	
		106	0.80	99.24	0.45-μm filter	
		106	33.0	68.87	2.5-μm filter	
		106	>100	<5.66	11.0-μm filter	
					100 mg/L Tartrate present:	
	10.0	106	0.16	99.85	0.025-μm filter	
		106	0.20	99.81	0.45-μm filter	
		106	16.5	84.43	2.5-μm filter	
		106	>100	<5.66	11.0-μm filter	
					No Chelants present:	
	10.0	106	>100	<5.66	5-min settling	
		106	4.2	96.04	15-min settling	
		106	3.6	96.60	30-min settling	

Table 4.2 cont.
Summary of Heavy Metal Removals Achieved Using Various Chemical Precipitation Techniques.

Metal	pH	Metal Concentration (mg/L) Initial	Residual	Removal Efficiency (%)	Comments	Reference
					100 mg/L Ammonia present:	
Zn	10.0	106	>100	<5.66	5-min settling	Ku & Peters 1986
		106	5.6	94.72	15-min settling	
		106	4.4	95.85	30-min settling	
					100 mg/L Citrate present:	
	10.0	106	>100	<5.66	5-min settling	
		106	>100	<5.66	15-min settling	
		106	>100	<5.66	30-min settling	
					100 mg/L Tartrate present:	
	10.0	106	>100	<5.66	5-min settling	
		106	>100	<5.66	15-min settling	
		106	>100	<5.66	30-min settling	
Zn	8.0	100	0.35	99.65	No chelants S^{-2} = 1.05x	Peters and Ku 1987
		100	0.60	99.40	23 mg/L Citrate S^{-2} = 1.05x	
		100	0.70	99.30	92 mg/L Citrate S^{-2} = 1.05x	
		100	0.85	99.15	225 mg/L Citrate S^{-2} = 1.05x	
		100	0.75	99.25	450 mg/L Citrate S^{-2} = 1.05x	
		100	0.80	99.20	900 mg/L Citrate S^{-2} = 1.05x	
Mixed Metals:	8.0					
Cd		2.06	0.10	95.1	Wastewater from electroplating and metal finishing operation at Ft. Belvoir, VA	Resta et al. 1978
Cr		2.61	0.32	87.7		
Cu		1.82	0.04	97.8		
Ni		3.50	0.07	98.0		
Zn		5.8	0.41	92.9		
Mixed Metals:						Fender et al. 1982
Zn	10.8	813	6.66	99.2	Pure hydroxide treatment	
Fe		<0.1	<0.1	—		
Pb		23.5	0.09	99.6		
Zn	11.1	813	6.74	99.2	50 mg/L FeS	
Fe		<0.1	<0.1	—		
Pb		23.5	<0.03	>99.8		
Zn	10.6	813	3.90	99.5	100 mg/L FeS	
Fe		<0.1	<0.1	—		
Pb		23.5	0.03	99.8		
Zn	10.4	813	4.59	99.4	500 mg/L FeS	
Fe		<0.1	<0.1	—		
Pb		23.5	0.10	99.6		
Zn	10.85	813	4.40	99.45	1000 mg/L FeS	
Fe		<0.1	<0.1	—		
Pb		23.5	0.03	99.8		
Zn	10.7	813	4.64	99.4	2000 mg/L FeS	
Fe		<0.1	<0.1	—		
Pb		23.5	0.04	99.8		

Table 4.2 cont.
Summary of Heavy Metal Removals Achieved Using Various Chemical Precipitation Techniques.

Metal	pH	Metal Concentration (mg/L) Initial	Residual	Removal Efficiency (%)	Comments	Reference
Mixed Metals:	8.0					
Zn		30-60	—	51-75	S^{-2} = 0.6x, full scale plant	Bhattacharyya et al. 1981a
Pb		20-40	—	97-99		
Cd		3-16	—	99		
Cu		3-5	—	90		
Hg		2-4	—	99		
Fe		5-20	—	2-26		
Mixed Metals:	8.5					Bhattacharyya et al. 1979
Cd		10.5	0.6	99.4	Copper smelting plant scrubber wastewater; hydroxide treatment only	
Cu		297	0.5	99.8		
Zn		85.5	3.1	96.4		
Pb		39.0	0.7	98.2		
Fe		149	<0.5	>99.7		
Se		3.0	<1.0	>66.7		
As		100	6.8	93.2		
Cd	8.0.-8.5	10.5	<0.05	>99.5	S^{-2} = 5 mM	
Cu		297	0.2	99.9		
Zn		85.5	0.9	98.9		
Pb		39.0	0.4	99.0		
Fe		149	<0.5	>99.7		
Se		3.0	<1.0	>66.7		
As		100	2.0	98.0		
Cd	8.0-8.5	10.5	<0.05	>99.5	S^{-2} = 8 mM	
Cu		297	0.1	>99.9		
Zn		85.5	0.05	99.4		
Pb		39.0	0.4	99.0		
Fe		149	<0.5	>99.7		
Se		3.0	1.2	60.0		
As		100	11.0	89.0		
Mixed Metals:						
Cd	8.0-8.5	10.5	<0.05	>99.5	S^{-2} = 12 mM	Bhattacharyya et al. 1981b
Cu		297	<0.05	>99.9		
Zn		85.5	<0.05	>99.5		
Pb		39.0	0.3	99.2		
Fe		149	<0.5	>99.7		
Se		3.0	1.0	66.7		
As		100	21.0	79.0		
Mixed Metals:						
Cu	8.0	100	0.4	99.6	No EDTA, S^{-2} = 1.05x	Peters et al. 1984b
Zn		100	0.3	99.7		
Cu	8.0	100	0.3	99.7	100 mg/L EDTA, S^{-2} = 1.05x	
Zn		100	15.0	85.0		
Cu	8.0	100	0.4	99.6	200 mg/L EDTA, S^{-2} = 1.05x	
Zn		100	27.0	73.0		
Cu	8.0	100	0.5	99.5	200 mg/L EDTA, S^{-2} = 0.8x	
Zn		100	74.0	26.0		

Table 4.2 cont.
Summary of Heavy Metal Removals Achieved Using Various Chemical Precipitation Techniques.

Metal	pH	Metal Concentration (mg/L) Initial	Residual	Removal Efficiency (%)	Comments	Reference
Mixed Metals:	9.0					
Cd		7.95	<0.05	>99.4	CaS precipitation	Kim 1981
Cu		18.6	<0.05	>99.7		
Cr		1.34	<0.05	>96.2		
Pb		3.5	<0.5	>85.7		
Zn		47.0	<0.05	>98.9		
Mixed Metals:						
Zn	8.0	24.5	2.4	90.2	Industrial plating wastewater; hydroxide treatment only	Peters and Ku 1984
Cu		62.4	1.45	97.7		
Cr		28.0	13.6	51.4		
Ni		1.04	0.36	65.4		
Zn	8.0	24.5	1.70	93.0	S^{-2} = 0.83x	
Cu		62.4	1.325	97.9		
Cr		28.0	8.8	68.6		
Ni		1.04	0.30	71.1		
Zn	8.0	24.5	0.40	98.4	S^{-2} = 1.0x	
Cu		62.4	1.40	97.75		
Cr		28.0	9.6	65.7		
Ni		1.04	0.30	71.1		
Zn	8.0	24.5	0.39	98.4	S^{-2} = 1.1x	
Cu		62.4	1.475	97.6		
Cr		28.0	8.0	71.4		
Ni		1.04	0.20	80.8		
Zn	10.0	24.5	0.5	97.9	Hydroxide treatment only	
Cu		62.4	14.4	76.9		
Cr		28.0	8.0	91.6		
Ni		1.04	0.15	85.5		
Zn	10.0	24.5	0.53	97.8	S^{-2} = 0.83x	
Cu		62.4	12.2	80.4		
Cr		28.0	2.05	92.7		
Ni		1.04	0.175	83.1		
Zn	10.0	24.5	0.605	97.5	S^{-2} = 1.0x	
Cu		62.4	12.4	80.1		
Cr		28.0	1.75	93.8		
Ni		1.04	0.175	83.1		
Zn	10.0	24.5	1.2	95.1	S^{-2} = 1.1x	
Cu		62.4	10.8	82.7		
Cr		28.0	1.50	94.6		
Ni		1.04	0.20	80.8		
Mixed Metals:	8.0-8.4					
Cd		3.3	0.06	98.2	FeS Dose = 1.5x	Brantner & Cichon 1981
Cr		0.52	<0.05	>98.2		
Cu		0.35	<0.03	>91.4		
Pb		4.5	<0.1	>97.7		
Zn		93.0	0.68	99.3		

Table 4.2 cont.

Summary of Heavy Metal Removals Achieved Using Various Chemical Precipitation Techniques.

Metal	pH	Metal Concentration (mg/L) Initial	Residual	Removal Efficiency (%)	Comments	Reference
Combined Precipitation Treatment						
Ni	6	10	8.2	18	Hydroxide treatment only	McAnally et al. 1984
	7	10	9.5	5		
	8	10	6.8	32		
	9	10	4.0	60		
	10	10	0.2	98		
	11	10	0.1	99		
	6	10	7.3	27	OH treatment, C_T = 50 mg/L	
	7	10	7.8	22		
	8	10	8.3	17		
	9	10	1.6	84		
	10	10	0.05	99.5		
	11	10	0.1	99		
	6	10	9.8	2	OH treatment, C_T = 100 mg/L	
	7	10	8.8	12		
	8	10	8.9	11		
	9	10	2.3	77		
	10	10	0.12	98.8		
	11	10	0.05	99.5		
	6	10	9.7	3	OH treatment, C_T = 200 mg/L	
	7	10	9.6	4		
	8	10	8.7	13		
	9	10	6.4	36		
	10	10	0.1	99		
	11	10	0.05	99.5		
	6	10	7.0	30	S_T=5 mg/L, C_T=0	
	7	10	9.4	6		
	8	10	7.8	22		
	9	10	1.5	85		
	10	10	0.1	99		
	11	10	0.05	99.5		
	6	10	8.7	13	S_T=5mg/L, C_T=50mg/L	
	7	10	8.4	16		
	8	10	8.8	12		
	9	10	3.2	68		
	10	10	0.3	97		
	11	10	0.05	99.5		
	6	10	10.0	0	S_T=5 mg/L, C_T=100mg/L	
	7	10	9.5	5		
	8	10	9.2	8		
	9	10	4.8	52		
	10	10	0.1	99		
	11	10	0.05	99.5		

Table 4.2 cont.
Summary of Heavy Metal Removals Achieved Using Various Chemical Precipitation Techniques.

Metal	pH	Metal Concentration (mg/L) Initial	Residual	Removal Efficiency (%)	Comments	Reference
Ni	6	10	9.2	8	S_T=5 mg/L, C_T=200mg/L	McAnally et al. 1984
	7	10	9.4	6		
	8	10	9.1	9		
	9	10	4.8	52		
	10	10	0.15	98.5		
	11	10	0.05	99.5		
	6	10	9.1	9	S_T=10 mg/L, C_T=0	
	7	10	7.6	24		
	8	10	7.7	23		
	9	10	4.6	52		
	10	10	0.1	99		
	11	10	0.05	99.5		
	6	10	8.9	11	S_T=10 mg/L, C_T=50 mg/L	
	7	10	7.8	22		
	8	10	8.1	19		
	9	10	4.5	55		
	10	10	0.1	99		
	11	10	0.05	99.5		
	6	10	9.1	9	S_T=10 mg/L, C_T=100 mg/L	
	7	10	9.2	8		
	8	10	8.6	14		
	9	10	6.8	32		
	10	10	0.9	91		
	11	10	0.05	99.5		
	6	10	9.2	8	S_T=10 mg/L, C_T=200 mg/L	
	7	10	8.8	12		
	8	10	8.5	15		
	9	10	6.6	34		
	10	10	0.1	99		
	11	10	0.05	99.5		
	6	10	9.7	3	S_T=20 mg/L, C_T=0	
	7	10	8.2	18		
	8	10	1.0	90		
	9	10	3.1	69		
	10	10	0.1	99		
	11	10	0.05	99.5		
	6	10	9.3	7	S_T=20 mg/L, C_T=50 mg/L	
	7	10	6.1	39		
	8	10	5.2	48		
	9	10	2.8	72		
	10	10	0.15	98.5		
	11	10	0.1	99		
	6	10	9.5	5	S_T=20 mg/L, C_T=100 mg/L	
	7	10	9.2	8		
	8	10	9.1	9		
	9	10	3.4	66		
	10	10	0.15	98.5		
	11	10	0.05	99.5		

Table 4.2 cont.

Summary of Heavy Metal Removals Achieved Using Various Chemical Precipitation Techniques.

Metal	pH	Metal Concentration (mg/L) Initial	Residual	Removal Efficiency (%)	Comments	Reference
Ni	6	10	8.5	15	S_T=20 mg/L, C_T=200 mg/L	McAnally et al. 1984
	7	10	9.0	10		
	8	10	8.8	12		
	9	10	5.4	46		
	10	10	0.3	97		
	11	10	0.05	99.5		
Ni	6	10	8.2	18	C_T = 0, Fe/Ni = 0	McFadden et al., 1985
		10	8.2	18	C_T = 0,Fe/Ni=0.5	
		10	4.9	51	C_T = 0, Fe/Ni=1.0	
		10	6.7	33	C_T = 0, Fe/Ni=2.0	
	6	10	7.6	24	C_T=50 mg/L, Fe/Ni=0	
		10	9.2	8	C_T=50 mg/L, Fe/Ni=0.5	
		10	9.5	5	C_T=50 mg/L, Fe/Ni=1.0	
		10	8.8	12	C_T=50 mg/L, Fe/Ni=2.0	
	6	10	9.9	1	C_T=100 mg/L, Fe/Ni=0	
		10	9.6	4	C_T=100 mg/L, Fe/Ni=0.5	
		10	9.3	7	C_T=100 mg/L, Fe/Ni=1.0	
		10	8.9	11	C_T=100 mg/L, Fe/Ni=2.0	
	6	10	9.5	5	C_T=200 mg/L, Fe/Ni=0	
		10	9.0	10	C_T=200 mg/L, Fe/Ni=0.5	
		10	8.8	12	C_T=200 mg/L, Fe/Ni=1.0	
		10	8.8	12	C_T=200 mg/L, Fe/Ni=2.0	
	9	10	7.7	23	C_T = 0, Fe/Ni = 0	
		10	0.5	95	C_T = 0, Fe/Ni=0.5	
		10	0.35	96.5	C_T = 0, Fe/Ni=1.0	
		10	0.35	96.5	C_T = 0, Fe/Ni=2.0	
	9	10	2.0	80	C_T=50 mg/L, Fe/Ni=0	
		10	1.7	83	C_T=50 mg/L, Fe/Ni=0.5	
		10	0.2	98	C_T=50 mg/L, Fe/Ni=1.0	
		10	0.3	97	C_T=50 mg/L, Fe/Ni=2.0	
	9	10	2.8	72	C_T=100 mg/L, Fe/Ni=0	
		10	1.9	81	C_T=100 mg/L, Fe/Ni=0.5	
		10	0.3	97	C_T=100 mg/L, Fe/Ni=1.0	
		10	0.22	97.8	C_T=100 mg/L, Fe/Ni=2.0	
	9	10	6.8	32	C_T=200 mg/L, Fe/Ni=0	
		10	2.0	80	C_T=200 mg/L, Fe/Ni=0.5	
		10	0.12	98.8	C_T=200 mg/L, Fe/Ni=1.0	
		10	0.20	98	C_T=200 mg/L, Fe/Ni=2.0	
	10	10	0.20	98	C_T = 0, Fe/Ni = 0	
		10	0.10	99	C_T = 0, Fe.Ni=0.5	
		10	<0.10	>99	C_T = 0, Fe/Ni=1.0	
		10	<0.10	>99	C_T = 0, Fe/Ni=2.0	

Table 4.2 cont.
Summary of Heavy Metal Removals Achieved Using Various Chemical Precipitation Techniques.

Metal	pH	Metal Concentration (mg/L) Initial	Residual	Removal Efficiency (%)	Comments	Reference
Ni	10	10	<0.10	>99	C_T= 50 mg/L, Fe/Ni=0	McFadden et al. 1985
		10	<0.10	>99	C_T= 50 mg/L, Fe/Ni=0.5	
		10	<0.10	>99	C_T= 50 mg/L, Fe/Ni=1.0	
		10	<0.10	>99	C_T= 50 mg/L, Fe/Ni=2.0	
	10	10	0.14	98.6	C_T=100 mg/L, Fe/Ni=0	
		10	0.10	99	C_T=100 mg/L, Fe/Ni=0.5	
		10	<0.10	>99	C_T=100 mg/L, Fe/Ni=1.0	
		10	<0.10	>99	C_T=100 mg/L, Fe/Ni=2.0	
	10	10	<0.10	>99	C_T=200 mg/L, Fe/Ni=0	
		10	<0.10	>99	C_T=200 mg/L, Fe/Ni=0.5	
		10	<0.10	>99	C_T=200 mg/L, Fe/Ni=1.0	
		10	<0.10	>99	C_T=200 mg/L, Fe/Ni=2.0	
Ni	10.0	5.0	0.35	93	$FeSO_4$ = 20 mg/L	Maruyama et al., 1975
Cd	6.90	1.0	0.08 Ca=147.5	92	Ca_i=150 mg/L as $CaCO_3$	Chang & Peters, 1985
	7.32	1.0	<0.01 Ca=129.5	>99		
	7.60	1.0	<0.01 Ca=119.4	>99		
	7.76	1.0	<0.01 Ca=75.0	>99		
	8.04	1.0	ND Ca=39.0	>99		
	9.44	1.0	ND Ca= 4.6	>99		
	10.38	1.0	ND Ca= 3.0	>99		
	10.84	1.0	ND Ca= 3.1	>99		
	7.42	1.0	<0.01 Ca=152.3	>99	Ca_i=250 mg/L as $CaCO_3$	
	7.45	1.0	<0.01 Ca=125.9	>99		
	7.66	1.0	<0.01 Ca= 92.5	>99		
	7.96	1.0	ND Ca= 51.7	>99		
	8.66	1.0	ND Ca= 11.5	>99		
	9.74	1.0	ND Ca = 4.2	>99		
	10.62	1.0	ND Ca = 3.7	>99		
	11.01	1.0	ND Ca = 3.7	>99		

Table 4.2 cont.

Summary of Heavy Metal Removals Achieved Using Various Chemical Precipitation Techniques.

Metal	pH	Metal Concentration (mg/L) Initial	Residual	Removal Efficiency (%)	Comments	Reference
Cd	7.36	1.0	<0.01; Ca=191.6	>99	Ca_i=350 mg/L as $CaCO_3$	Chang & Peters 1985
	7.41	1.0	ND Ca=156.7	>99		
	7.57	1.0	ND Ca=117.0	>99		
	7.62	1.0	ND Ca= 92.1	>99		
	7.64	1.0	ND Ca=106.5	>99		
	8.03	1.0	ND Ca= 37.1	>99		
	9.58	1.0	ND Ca= 10.3	>99		
	10.81	1.0	ND Ca = 2.7	>99		
	7.79	5.0	0.02 Ca=148.9	98	Ca_i=150 mg/L as $CaCO_3$	
	8.05	5.0	0.01 Ca=110.0	99		
	9.42	5.0	<0.01 Ca= 38.7	>99		
	10.96	5.0	ND Ca = 52.2	>99		
	11.30	5.0	<0.01 Ca= 32.2	>99		
	7.82	5.0	0.02 Ca=115.8	98	Ca_i=250 mg/L as $CaCO_3$	
	8.21	5.0	0.02 Ca= 75.0	98		
	8.82	5.0	<0.01 Ca= 40.5	>99		
	10.47	5.0	ND Ca = 11.2	>99		
	11.07	5.0	ND Ca = 14.4	>99		
	11.39	5.0	0.02 Ca=10.6	98		

Table 4.2 cont.
Summary of Heavy Metal Removals Achieved Using Various Chemical Precipitation Techniques.

Metal	pH	Metal Concentration (mg/L) Initial	Residual	Removal Efficiency (%)	Comments	Reference
Cd	7.67	5.0	0.03 Ca=208.1	97	Ca_i=350 mg/L as $CaCO_3$	Chang & Peters, 1985
	7.97	5.0	<0.01 Ca=119.4	>99		
	8.38	5.0	ND Ca = 67.2	>99		
	10.09	5.0	<0.01 Ca= 22.8	>99		
	10.66	5.0	0.01 Ca= 14.8	99		
	10.95	5.0	<0.01 Ca= 15.3	>99		
	11.30	5.0	ND Ca = 10.8	>99		
Mixed Metals:						
Zn	8.0	50	0.57	98.9		Talbot 1985
Ni		15	1.8	88.0		
Pb		15	<0.05	>99.7		
Cd		15	<0.05	>99.7		
Cu		15	<0.03	>99.8		
Hg		2.9	<0.001	>99.9		
Zn		50	0.04	99.9		
Ni		15	<0.05	>99.7		
Pb		15	<0.05	>99.7		
Cd		15	<0.05	>99.7		
Cu		15	<0.03	>99.8		
Hg		2.9	<0.001	>99.9		

5 PROCESS EVALUATION

Both oxidation and precipitation processes appear to be excellent choices for use in the treatment of groundwaters containing the types of contaminants discussed in Chapter 4.0. Systems are commercially available for drinking water and wastewater treatment. The need, therefore, is for transfer of existing technology to another application, rather than development of a new technology.

For over a decade, substitution processes have been available for the treatment of soils and sludges contaminated with polychlorinated biphenyls (PCBs) and other chlorinated organics. They have not been used extensively because other methods and technologies, such as landfilling and incineration, have been cheaper and more readily available. Where local community concerns prevented the use of existing technologies, the "do nothing" alternative often has been employed. This situation has resulted in very little economic incentive to invest in the development of these technologies. A number of substitution processes now appear to be available at costs approaching those of incineration, specifically, Galson Research Corporation (GRC), SoilTech/Anaerobic Thermal Processor (ATP), and KGME/DECHLOR coupled with a thermal treatment system for soils.

The Base Catalyzed Decomposition (BCD) Process, which appears to be similar to the SoilTech/ATP Process, but operates at lower temperatures and uses an improved reagent, is an emerging technology. It is discussed in Appendix A.

6 LIMITATIONS

Chemical treatment processes are highly specialized. The selected process must not only be effective in treating the particular contaminants in a particular matrix, but it must also be chemically compatible with the other constituents of the matrix.

6.1 Substitution Processes

Substitution processes are applicable only in treating substituted organic contaminants, such as halogenated or sulfur bearing organics. In fact, they have been successfully applied only in the treatment of chlorinated aromatic compounds, such as polychlorinated biphenyls (PCBs) and chlorodibenzodioxins. The processes are further limited when water is present. In the treatment of contaminated soils, the processes described herein can tolerate some water, but water increases the reagent requirements. In high temperature substitution processes, water increases the removal of contaminants by competing mechanisms, such as steam stripping, and increases fuel requirements.

6.2 Oxidation Processes

Advanced oxidation processes were developed in order to overcome many of the limitations encountered in general oxidation processes. Since this technology is based on hydroxyl free radical chemistry, chemical interactions are highly nonspecific and nonselective. Because of the high reactivity of hydroxyl radicals, most contaminants in most aqueous streams are very effectively destroyed. The matrix limitations encountered in using these technologies are similar to those encountered in devising strategies based on hydroxyl radical chemistry. The biggest concern is the presence of scavengers, such as bicarbonate and carbonate ions. At high concentrations, higher doses of ozone, hydrogen peroxide, and ultraviolet (UV) radiation are required. Another important factor is penetration of UV light through the wastewater stream. Light penetration is attenuated by high particle concentrations, and, therefore, the technique generally is not well suited to treating soils. Rates of destruction will vary with such factors as the nature of the contaminant mixture, pH, concentration of contaminants, presence of scavengers (carbonates and natural organic material), and the inorganics present.

6.2.1 UV-Ozonation — Treatment of Water

Depending on the reaction conditions (pH, ions, temperature, etc.), organic compounds interact with ozone via multiple pathways. Ionic strength and pH do not affect gas transfer, although the solubility of ozone increases with decreasing temperature. The kinetic regime of ozone reactions is highly dependent on pH and ozone partial pressure. For example, the degradation kinetics of p-nitrophenol are very slow at pH 2, but they increase with pH until an instantaneous rate is achieved at pH 8.5 (Beltran, Gomez-Serrano, and Duran 1992). A similar trend is observed with increasing ozone partial pressure. These results have important implications for design of ozone contactors because, under extreme conditions, mass transfer of ozone can be controlling and interfacial areas should be maximized. At intermediate pH, rates of mass transfer and chemical reaction are about equal. For nitrophenols, ozone decomposition or free radical pathways become major only above a pH of 12.4. Therefore, to select this pathway, ozone decomposition requires catalysis by either UV light or hydrogen peroxide. In general, high pH favors ozone decomposition and has a favorable effect also on substrate ionization by causing deprotonation, which produces either uncharged or negatively-charged species.

A major consideration in applying ozonation in remediation is the presence of competitive substrates and inhibitors or scavengers. For groundwater in particular, the inorganic matrix can have a profound effect on ozone efficiency. Bicarbonate and carbonate ions scavenge hydroxyl radicals, terminating the decomposition chain reaction and stabilizing molecular ozone. Once again, advanced oxidation processes are usually required to offset this effect. Some ions, iron(II) for example, will enhance ozone kinetics by initiating decomposition.

The natural organic matter (NOM) exerts very complex effects on ozonation efficiency. This class of ill-defined compounds can participate in initiation, propagation, and termination interactions. The extent of these interactions will vary regionally and seasonally. Moreover, NOM is usually present at much higher concentrations than synthetic organic pollutants. In general, micropollutant destruction in the presence of even low to moderate amounts of NOM requires relatively high ozone doses, and AOPs would typically be used.

6.2.2 UV-Oxidation — Treatment of Particulate

Degradation of polyaromatic hydrocarbons (PAHs) adsorbed onto a silica gel was studied in a fluidized-bed reactor simulating atmospheric interactions (Alebic-Juretic, Cuitas, and Klasinc 1990). Rates of degradation were influenced by the degree of surface coverage and enhanced on the surface of the acidic silica gel. These findings suggest that the type of surface is important and that reaction is via electrophilic substitution. Ozonation in heterogeneous aqueous systems has not been widely studied and it is expected that ozone effectiveness would be diminished by the presence of solids because of a quenching effect and mass transfer limitations. A major difference between atmospheric and aqueous particles lies in the higher concentrations of NOM and other potential scavengers in aqueous systems.

Although UV radiation does not penetrate solid matrices such as soils, direct or indirect photolysis can be used effectively in soil systems. Shallow depths and periodic mixing are needed with soil irradiation.

6.2.3 UV-Hydrogen Peroxide — Treatment of Water

A number of water soluble, organic contaminants in groundwater have been successfully oxidated through the UV/H_2O_2 Process (Camp 1991; Edwards and Bonham 1988; Rowland 1989; US EPA 1992b). While the process appears to be effective in treating most of the contaminants in groundwater studied to date, there are a number of limitations that can affect its effectiveness in treating contaminated water.

Interference with transmission of light energy from the source (lamp) through the aqueous medium to the substrate being photolyzed can adversely affect the efficiency of the UV/H_2O_2 Process. Optical fouling of the quartz tubes containing the UV light source over the course of the treatment process can significantly reduce process performance and efficiency. An example of this difficulty has been described by Camp (1991) in the treatment of chlorinated hydrocarbons in groundwater containing less than a 0.1 mg/L of iron. Camp reported iron deposition rates of 0.009 to 0.085 mg Fe/cm^2hr in the treatment of groundwater containing tetrachloroethene (PCE). The resultant loss in light transmission translated to a reduction in overall process efficiency, which worsened over the course of several days. Removing the deposited iron from the quartz tubes improved process performance. The efficiency of PCE destruction decreased as the intervals between cleanings of the quartz tubes increased. The iron deposition process was thought to be an oxidative precipitation or flocculation of iron. Therefore, the presence of even relatively low concentrations of dissolved minerals may adversely impact light transmission and process efficiency. The quartz tubes containing the light source may need to be cleaned periodically, and the effect on operating costs, labor, maintenance, and downtime of the process, will need to be considered.

Dissolved carbonate (CO_3^{-2}) and bicarbonate (HCO_3^-) ions at relatively high concentrations (≈4 mmol/L) can significantly reduce the rate of degradation of organic contaminants (Guittonneau et al. 1990). Both carbonate and bicarbonate ions react with hydroxyl radicals as scavengers with bimolecular rate constants in the range of $1x10^{7-8}$ L $mol^{-1}S^{-1}$, which is competitive with the rate constants for the reaction of hydroxyl radicals with most organic substrates. Therefore, pH adjustments with carbonate bases prior to the oxidation reaction should be avoided. If necessary, sodium hydroxide (NaOH) should be used.

Since hydroxyl radicals are potent and nonselective oxidizing agents, the presence of other dissolved organics in water may require the use of higher levels of hydrogen peroxide beyond the amounts required to degrade the contaminants of interest.

Virtually all of the organic contaminants which undergo UV/H_2O_2 oxidation produce intermediate products that must also undergo subsequent oxidation. These by-products, depending upon their nature, may require extended oxidation and thus may impact reactor productivity by increasing the total residence time for complete mineralization.

These process limitations need to be addressed in treatability studies and matrix analysis and characterization to determine the suitability and cost of applying the UV/H_2O_2 oxidation process to a site remediation.

6.2.4 UV-Hydrogen Peroxide — Treatment of Soils and Sediments

Since photochemical reaction efficiency depends upon the optical path length of medium in which it is carried out, direct application of the UV/H_2O_2 oxidation process to soils is not practical. If, however, the contaminants of interest can be effectively transferred to an aqueous phase of reasonably high optical path length, the UV/H_2O_2 oxidation then can be effectively carried out on soils or sediments. This approach has been applied to lagoon sediments contaminated with water soluble explosives waste (Wentsel, Sommerer, and Kitchens 1981). The presence of high levels of water soluble organics, other than the contaminants to be degraded, can result in the need for higher levels of hydrogen peroxide than desired and, therefore, increase costs.

6.3 Precipitation Processes

Precipitation processes are limited to the treatment of inorganic materials in aqueous media.

7 TECHNOLOGY PROGNOSIS

Under proper conditions, discussed in this monograph, chemical treatment can be a useful site remediation technology. The following are likely applications of the processes addressed in this monograph:

- Substitution processes, especially the high temperature processes, will be used to treat soils and sludges contaminated with polychlorinated biphenyls (PCBs), pentachlorophenol (PCP), chlorodibenzodioxins, and chlorodibenzofurans;
- Oxidation and precipitation processes will be used to treat water from pump-and-treat applications; and
- Precipitation processes will be commonly used to treat sludges and aqueous streams that are contaminated with toxic metals and other cations.

APPENDIX A

Emerging Technologies

There are a number of chemical technologies that are at a research or an early developmental stage. Although there are some questions about their economic feasibility, they appear to be very promising technologically. Six such technologies are briefly discussed below.

A.1 *The Base Catalyzed Decomposition (BCD) Process*

The BCD, a chemical reaction method similar in many ways to the anaerobic thermal processor (ATP) process, is under development by the Risk Reduction Engineering Laboratory (RREL) at the U.S. Environmental Protection Laboratory, Cincinnati, Ohio. This process has been tested in the laboratory for treating a variety of contaminants including polychlorinated biphenyls (PCBs); 2,4-D and 2,4,5-T (see table A.1 on page A.2); tetrachlorodibenzo-p-dioxin (TCDD), tetrachlorodibenzofurans (TCDF), lindane, endrin, dieldrin, and other chlorinated compounds. (Kornel, Rogers, and Sparks 1991a, 1991b, and 1991c).

In these tests, 10 mL of oil (Sun Par LW 107 boiling range: 327° to 387°C) contaminated with 2,4-D, and 2,4,5-T were combined with 1.0 g NaOH and 1.0g of catalyst in a 50 mL round bottom flask. A vertical condenser was attached to the reaction flask, and the mixture was heated and maintained at 330 to 340°C (630 to 640°F) for one hour under conditions of refluxing solvents (Kornel, Rogers, and Sparks 1991a, 1991b, and 1991c). After three hours, the mixture was cooled and 50 mL of distilled water was added to dissolve sodium chloride formed during the reaction. Water ex-

Table A.1
BCD Treatment of 2,4-D; 2,4,5-T

Herbicide	Weight Of Sample (mg)	Theoretical	Actual	% Chlorine Balance
2,4-D	1,005	324.6	373.7	115.0
2,4-D	1,009	325.7	335.4	102.9
2,4,5-T	1,004	420.0	421.1	100.3
2,4,5-T	1,018	427.6	450.7	105.4

TCDD in Liquid Herbicide Formulation

Tests	Ratio of Formulation:Oil	Treatment Time (hrs)	TCDD Concentration (ppb) Before **	After ***
2-1	1:3	4	11.3	0.13
2-2	1:3	4	11.3	0.017
2-3	110	3	4.8	0.005
2-4	2:0	83	8.2	0.01
2-5	1:3	8	11.3	0.01
2-6	1:1	3	18.3	0.002*

* Detection limit of high resolution MS
** Calculated in the initial reaction mixture
*** In the final reaction mixture

tract was filtered, diluted to 100 mL with distilled water and analyzed for chloride by ion chromatography. Experimental results shown in Table A.1 confirmed the conversion of the aromatic chloride to sodium chloride for 2,4-D and 2,4,5-T.

It was claimed that the BCD process totally dechlorinates 2,4-D and 2,4,5-T at 10% starting concentrations. The developers claimed that the reaction completely replaced the chlorine with hydrogen, but no independent verification of this phenomenon has been published.

The BCD chemistry was used to destroy up to 100,000 ppm of PCBs in dielectric fluids. The treatment was similar to the one described above. Gas chromatography showed an absence of biphenyl, and the presence of pentachloro-, hexachloro-, and heptachloro-biphenyl in the untreated mixture. Similar analysis of the reacted mixture showed the presence of biphenyl and absence of the chlorobiphenyls. Chlorine balance accounts for the chlorine present as the mineralized salt.

A.2 Iron (II) Catalyzed H_2O_2 Oxidation (Fenton's Reagent)

Iron II can potentially be used as a catalyst to improve the performance of hydrogen peroxide as an oxidizing agent. While no large-scale or pilot application of this technology has been found, there appears to be no chemical impediment to the procedure. This type of chemical oxidation is frequently used in the laboratory and the costs appear to make its application in remediation worth considering.

Among the most commonly used metal catalysts for the decomposition of hydrogen peroxide in aqueous media is iron (Fe^{2+}). This basic process was first reported by Fenton (1984) in describing the oxidation of tartaric acid. The combination of various iron salts with hydrogen peroxide has come to be known as Fenton's Reagent. The resultant oxidation process has also come to be referred to as the Fenton Process.

The catalytic reaction of iron with hydrogen peroxide in the Fenton Process is believed to proceed through a series of redox reactions (Watts, Tyre, and Miller 1991) represented by the following equations, which produce powerful oxidizing agents in the form of hydroxyl radicals (HO•) and perhydroxyl radicals (HO_2•).

$$Fe^{2} + H_2O_2 \rightarrow Fe^{3} + HO^{-} + HO\bullet \qquad [A.1]$$

$$Fe^{2} + HO^{\bullet} \rightarrow Fe^{3} + HO^{-} \qquad [A.2]$$

$$Fe^{3} + H_2O_2 \rightarrow Fe^{2} + H^{+} + HO_2\bullet \qquad [A.3]$$

$$Fe^{3} + HO^{\bullet}{}_2 \rightarrow Fe^{2+} + H^{+} + O_2 \qquad [A.4]$$

Intermediate oxidation by-products can be formed under Fenton Process conditions and these must be taken into account in the mass balance of the process and for their environmental impact.

Fenton's Reagent consists of the combination of hydrogen peroxide (H_2O_2) and iron(II) salts (Fe^{2+}). The most common source of ferrous ion used in the laboratory process is ferrous sulfate, typically obtained as its heptahydrate ($FeSO_4{\bullet}7H_2O$). Commercially, ferrous sulfate is available as a moist bulk product with a specified range of water content. Hydrogen peroxide is commercially available in a range of concentrations from 30 to 70% by weight in water. Typically, hydrogen peroxide is most easily and safely handled at concentrations of 10 to 30% by weight in water. Higher concentrations can react violently with organic material.

The Fenton oxidation is normally carried out by first establishing the ferrous ion concentration at the desired level in the reaction medium by the addition of a ferrous salt. The reaction medium is then adjusted to pH 2 to 5 by the addition of a suitable acid, such as hydrochloric acid (HCl), or a base, such as sodium hydroxide (NaOH). The reaction medium is then agitated vigorously while hydrogen peroxide is slowly added to the reaction (Walling 1975). The reaction is normally carried out at ambient temperatures. In most cases, the optimum initial pH is three, although Fenton's reaction has been observed to produce OH^- efficiently at neutral pH (Zepp, Faust, and Hoigne 1992). Furthermore, the authors of this study found that light accelerated the reaction by photo-regenerating Fe(II).

In a slight modification of the Fenton's reaction, the complete mineralization of phenoxyacetic herbicides was observed in an iron(III)/H_2O_2 system (Pignatello 1992).

The degradation reaction was sensitive to pH, reaching an optimum around pH 2.7. The reaction was adversely affected by the presence of methanol and chloride, which exerted a scavenging effect on the hydroxyl radical, and by sulfate, which complexed Fe^{3+}. In this case, too, the rate of degradation was increased significantly by light. A partial explanation of the observed phenomena was that photoreduction of the iron monohydroxo-complex ($FeOH_2^+$) produces Fe^{2+} and OH^-. The ferrous ion in the presence of H_2O_2 generated hydroxyl radical in the conventional Fenton's reaction. Photo-assisted mineralization of the herbicides was nearly complete in 2 hours at a H_2O_2:herbicide ratio of 5:1, suggesting that dioxygen was also involved in the degradation pathway.

During the course of the reaction, oxygen gas is liberated as the hydrogen peroxide undergoes decomposition. This gas evolution can become violent if the hydrogen peroxide is added too rapidly. The gas evolved during the addition of hydrogen peroxide can potentially entrain volatile organic contaminants and should therefore be scrubbed appropriately prior to discharge into the environment.

The specific concentrations of ferrous ion and hydrogen peroxide required for complete mineralization of various contaminants will be dependent upon a number of factors related to the specific compounds and matrix to be treated. In the treatment of contaminated aqueous streams, the total dissolved carbon loading will determine the amount of hydrogen peroxide required for treatment. This requirement is a direct reflection of the nonselective nature and high reactivity of the hydroxyl radicals generated in the Fenton Process. Thus, the target compounds of interest and any other dissolved carbon sources will all consume hydrogen peroxide in the process.

The concentration of ferrous ion used in this process is based on the amount of hydrogen peroxide required for complete oxidation. Typical ferrous ion to hydrogen peroxide concentration ratios fall in the range of 1:200 to 1:1,000 by weight. In adding ferrous salts to the reaction medium, one should take into account the concentration of indigenous iron species. In the direct treatment of soils, the presence of naturally occurring ores containing ferrous ions, such as magnetite, may be sufficient to eliminate the need for ferrous salt addition (Vasilenko and Fedosoua 1987). The time required to complete the process, again, depends upon the amount of contaminant and indigenous carbon to be oxidized, as well as the concentration of ferrous ion and hydrogen peroxide used. Typical reaction times fall in the range of one to several hours.

Upon completion of the oxidation process, excess hydrogen peroxide may require removal or destruction. Finally, the concentration of soluble iron species in the treated medium may require adjustment to acceptable discharge levels by sequestering or precipitating of soluble iron species.

The range of contaminants treatable by the Fenton oxidation process is, in principle, quite broad. Several classes of organic compounds have been degraded on a laboratory scale, including polyaromatic hydrocarbons (PAHs), pesticides, aromatic hydrocarbons, phenols, chlorinated aromatics, chlorinated hydrocarbons, and lignins. The complete mineralization of benzo(a)pyrene and phenanthrene has been reported using Fenton oxidation (Kelley, Gauger, and Srivastava 1990). Phenol degradation using Fenton's Reagent has been reported by Osaki, Sugihara, and T. Kaji (1990). Chlorobenzene and chlorophenol are degraded by Fenton's Reagent via a number of complicated pathways, the more direct involving hydroxylation, ring cleavage, and mineralization (Sedlak and Andren 1991). Oxygen was present, the more direct pathway was favored, and less H_2O_2 was required. Gold, Kutsuki, and Morgan (1983) reported the rapid degradation of ^{14}C-labeled lignins, and a similar polymer degradation has been investigated by Matsuzuru et al. (1982) using Fenton's Reagent to degrade powdered cation exchange resins. Watts, Tyre, and Miller (1991) studied the laboratory-scale degradation of Dieldrin, Trifluralin, pentachlorophenol, and n-hexadecane in the Fenton Process. In that study, Trifluralin and pentachlorophenol were rapidly degraded by the iron/hydrogen peroxide system, while dieldrin and n-hexadecane were degraded more slowly. Decolorization of a wide variety of dyes has been demonstrated, and the best results were found below pH 3.5 (Kuo 1992), at which about 90% of the chemical oxygen demand and about 97% of the color was removed.

Water. Fenton oxidation is a very effective process for the treatment of contaminated wastewater. The strong oxidizing conditions produced in the process generally lead to the mineralization of all soluble organic compounds. This means that nonhazardous, naturally occurring organics are degraded along with the contaminant of interest and will, therefore, consume additional reagents beyond the amount required to degrade the target compounds. Water containing high levels of soluble or suspended organic matter may require very large amounts of reagents. The presence of useful forms of dissolved iron can, in principle, be used to advantage in reducing the amount of ferrous salts required to carry out the process.

Soils. The degradation of contaminants in soil matrices has been demonstrated in laboratory and field studies. The limitations described for treatment of water are also applicable to soils. In addition, soils containing a high level of organic matter can lead not only to use of very large amounts of reagent, but also to slower rates of contaminant degradation.

Under conditions that lead to complete mineralization of the target contaminant, the by-products of the Fenton process should be water and stable oxides that pose no threat to the environment. Conditions leading to complete mineralization, however, will need to be established for each matrix and contaminant type to be treated. The process parameters of stoichiometry, pH, and reagent concentration will need to be determined in an appropriate treatability study. Under Fenton conditions, the partial oxidation of certain organics can lead to undesirable by-products. An example of this has been reported by Heindl and Hutzinger (1987), who found that partial Fenton oxidation of trichlorobenzenes can lead to the formation of polychlorodibenzofurans (PCDFs) and polychlorodibenzodioxins (PCDDs).

When contaminants are completely mineralized in the Fenton oxidation, the only posttreatment requirements would be the adjustment of the matrix pH, the removal of iron to meet discharge permit levels, and the removal of any excess hydrogen peroxide. Complete mineralization should be verified by mass balance and analysis prior to discharge.

There have been only a few instances of full-scale remediation using the Fenton oxidation. One example is found in EPA's computerized on-line information system (COLIS). In this case 6,400 kg (14,000 lb) of phenol leaked from a railcar in Charleston, South Carolina, in May 1978. Runoff from the leak was collected and transferred to two 500,000 gal elastomer-lined and covered concrete bunkers where 50% H_2O_2 was metered into the bunker and circulated with hydrated ferric sulfate. Phenol was oxidized within 2 days, and initial concentrations, which ranged from 700 to 2,700 ppm, were reduced to less than 5 ppm.

Tate (1991) reported a one day full-scale production test by Environmental Health Research and Testing (EHRT) of an oxidation process which was used to treat soil containing up to 20,000 ppm hydrocarbon contamination at the Hamilton Army Airfield in Novato, California. The results of the test demonstrated the reduction of hydrocarbons to below 100 ppm. Full-scale remediation of this 18,000 m^3 (24,000 yd^3) site using Fenton technology under the supervision of the US Army Corps of Engineers was recently confirmed by a representative of Environmental Health Safety and Testing (Purbaugh 1992). Production throughput, according to EHRT, was approximately 1,200 m^3 (1,500 yd^3) per day based on a one shift operation. The remediation required six months to complete at an estimated cost of \$157/m^3 (\$120/yd^3). The EHRT could not offer engineering details beyond those contained herein because of proprietary considerations.

Technical grade hydrogen peroxide (35% by weight in water) is commercially available in tank car volumes at a cost of approximately $0.55/kg ($0.25/lb), and ferrous sulfate is available in moist bulk forms at a cost of $37/tonne ($34.00/ton). For matrices containing low levels of hazardous organic compounds and other organic material, the cost of reagents required to treat the matrix should be in the range of $22 to $55/tonne ($20 to $50/ton). Some field work using Fenton's reagent has been conducted by the

Institute of Gas Technology in Chicago, IL. This work came to the authors' attention after the monograph was written and was not included.

A.3 Photocatalysis in Semiconductor Systems

Semiconductors are materials that undergo a charge separation when irradiated with sufficiently energetic light. Electrons in the valence band of a semiconductor particle are excited to the conduction band by the absorption of a quantum of ultraband gap light, leaving behind a positive hole. These charges may recombine or migrate to the surface of the particle where they can engage in interfacial redox reactions. This phenomenon has been extensively applied in solar power generation and organic synthesis, as well as in other industrial uses (Heller 1987). Only relatively recently, however, has it come to the attention of environmental engineers. It has been demonstrated that photocatalysis can reduce metals or oxidize (in many cases, completely mineralize) almost every class of organic compound (Schiavello 1988; Ollis 1985; Ollis, Pelizzetti, and Serpone 1989; Pelizzetti, Minero and Maurino 1990). A novel photocatalytic method was recently reported for detoxifying cyanide wastes (Bhakta et al. 1992).

Some common examples of semiconductors are titanium dioxide (TiO_2), zinc oxide (ZnO), and cadmium sulfide (CdS). The anatase form of TiO_2 has received the most attention in the treatment of low levels of contaminants in either liquid or gaseous phases, because it appears to be more reactive and is commercially available at low cost (i.e. Degussa). Ultraviolet light having wavelengths less than 380 nm is required to induce the charge separation in TiO_2, and it has been demonstrated that solar radiation, the spectrum of which is 5% ultraviolet (UV), can drive these degradation reactions (Matthews 1987; Pilizzetti et al. 1988; Gerischer and Heller 1992; Manilal et al. 1992).

There is extensive literature on photocatalysis, and the theoretical basis of this process is well-established. It is interesting, however, that the exact mechanism by which organic compounds are destroyed remains unknown. It is conventional thinking that oxygen is required to sweep trapped electrons on the surface, inhibiting charge recombination and allowing oxidation to occur at the trapped positive hole. Oxidation may result from the transfer of an electron from an adsorbed organic compound, or it may be mediated by the hydroxyl free radical produced by the oxidation of water. Most kinetic treatments of photocatalysis are based on the assumption that the hydroxyl free radical is the primary oxidant and utilize Langmuir-Hinshelwood rate forms (Matthews 1988; Turchi and Ollis 1990). The rate-limiting step has been suggested to be electron scavenging by oxygen (Gerischer and Heller 1991) and some recent reports suggest that oxygen plays a more intrinsic role in the degradation processes than simply electron sweeping (Stafford et al. 1993; Barreto, Gray, and Anders 1994).

Most investigations of semiconductor photocatalysis have been conducted with single components in TiO_2 slurry systems and under batch laboratory conditions. Typically, a mercury vapor lamp is used to induce the charge separation of the semiconductor. An alternative process has been demonstrated whereby a colored pollutant can sensitize semiconductors by injecting electrons with visible light excitation (Dieckmann, Gray, and Kamat 1992 a, b). Application of photocatalysis necessitates the development of a reactor utilizing immobilized semiconductors, and TiO_2 has successfully been attached to glass mesh, to the inside of a glass coil, and to alumina (Matthews 1987; Magrini et al. 1992). A patented Nulite photoreactor uses stationary TiO_2 attached to a fiberglass mesh (Al-Ekabi et al. 1991).

Groundwaters or waters having inorganic ions or metals present can be treated effectively with photocatalysis, although rates may be adversely affected, and some loss of catalytic activity with time has been observed (Abdullah, Low, and Matthews 1990; Magrini et al. 1992). Using a pilot-scale, continuous flow slurry reactor that is powered by a parabolic solar collector, trichlorethylene (TCE) contamination in a groundwater located at a Superfund site in Livermore, California, has been reduced from 5,000 ppb to the detection limit of 5 ppb (U.S. Water News 1991, 12). Similarly, favorable reports have been made for the treatment of volatile organic compounds (VOCs) in the gas phase (Environment Today 1992, 10-11). Suspended solids may affect reaction rates by attenuating light penetration into the reactor and by partitioning contaminants. A preliminary comparison of projected operating costs among activated carbon, UV/ozone and UV/photocatalysis found that, for medium to high flows, photocatalysis is competitive with carbon and appears to be more economical than UV/ozone (Ollis et al. 1989). A more recent cost assessment based on the Livermore pilot study found that photolysis is more cost-effective than carbon, but that catalyst improvements are required for it to be competitive with UV/peroxide (Turchi 1992). Actual operating costs for the Nutech reactor are claimed to be very reasonable at \$0.24/m^3 (\$0.18/yd^3).

A.4 ***Ionizing Radiation***

A recent innovation in the treatment of industrial and municipal wastes is the use of high-energy radiation, which interacts with matter to produce ions, free radicals, and other short-lived reactive species (Singh, Kremers et al. 1985; Singh, Sagert et al. 1985). In an aqueous system, irradiation with ^{60}Co produces the ionizing event, forming, reducing (the aqueous electron) and oxidizing (the hydroxyl free radical) species in equal proportions. These radicals can then interact with contaminants and induce the same kind of reactions that have been discussed in the text of this monograph (see, for example, Singh et al. 1985; Kurucz et al. 1990; Nickelsen et al. 1992; Farooq et al. 1993). There also appear to be some advantages for the use of radiolysis in treating compounds in an adsorbed state (Dickson and Singh 1986). A pilot-scale study has been conducted over the last five years in Miami, Florida, by researchers at the University of Miami and Florida International University with funding from the National Science Foundation. This group has demonstrated that hazardous organic compounds can be effectively removed from aqueous solution economically, although treatment costs are highly dependent on the required dose and flow rate (Kurucz et al. 1990). Pulsed electron beams, produced by compact and inexpensive induction accelerators, appear to be a cost-effective means for removing a broad spectrum of toxic organic contaminants, and new accelerator technology is making this a very competitive technology (Science Research Laboratory, Inc., Somerville, Mass.).

Currently, radiolytic processes are used extensively in the curing of plastics where ionizing radiation replaces thermal or catalytic techniques and results in lower emissions. This is an excellent example of waste minimization. In general, capital and operating costs of electron beam processing are highly variable, but capital costs are normally high for conventional electron beams, and processing rates must be high to achieve reasonable unit costs (Cleland 1992).

A.5 *Sonication*

Ultrasonic waves in liquids can produce and accelerate many chemical reactions inside, at the interface, or at some distance from cavitating gas bubbles. High temperatures and pressures exist inside a collapsing bubble and at the interface. Hydroxyl hydrogen free radicals are produced in the thermal decomposition of water. Pyrolysis and radical reactions can occur simultaneously and appear to be responsible for the destruction of p-nitrophenol in aqueous solution (Kotronarou, Mills, and Hoffmann 1991). The destruction of parathion by ultrasonic irradiation has been reported also (Kotronarou, Mills, and Hoffmann 1992). The characteristics of pentachlorophenate degradation by means of sonochemical treatment at 530 kHz has been explained recently (Petrier et al. 1992). SRI International of Menlo Park, California, is in the process of scaling up a sonication process to enhance the photocatalysis (Environment Today 1992, 10-11). In this application, it is thought that ultrasound enhances primarily mass transfer, and a commercial sonication reactor would range from 10 to 50 kHz. A pilot-scale system is under development and is funded by DOE's Innovative Technologies Program.

A.6 *Iron(VI)-Ferrate*

Iron(VI)-Ferrate (not to be confused with Fenton's Reagent) is a high oxidation state of iron that is present in water as a divalent anion, FeO_4^{2-} and can be obtained commercially as a potassium salt under the trade name, TRU/Clear from Analytical Development Corporation (Colorado Springs, CO). Theoretically, ferrate possesses the properties of a strong oxidant (see table A.2 on page A.12), as well as the typical complexation and precipitation properties associated with iron(III). There has been limited research on the use of this chemical in water and wastewater treatment as a disinfectant, oxidant, and coagulant. For the most part, its use has remained of academic interest because, at a cost of $56/kg ($25/lb), it is extremely expensive, and until recently, it had not been found to be superior to other more affordable oxidation/precipitation strategies. Although the mechanism is not known, ferrate has been shown to produce better removal rates and less sludge in treating of waters contaminated with radioactive elements, especially the

Table A.2
Relative Oxidizing Strength of Oxidants

Oxidant	Chemical/ Species	Oxidation Potential* (Volts)	Relative Oxidation Strength
Fluorine	F_2	3.06	2.25
Hydroxy Radical	HO•	2.80	2.05
Atomic Oxygen	O•	2.42	1.78
Ferrate	FeO_4^{2-}	2.20	1.62
Ozone	O_3	2.07	1.52
Hydrogen Peroxide	H_2O_2	1.77	1.30
Perhydroxyl Radical	HCO•	1.70	1.25
Hypochlorous Acid	HOCl	1.49	1.1
Chlorine	Cl_2	1.36	1.00

*Source: Rice 1981

transuranic elements. A bibliography of work utilizing potassium ferrate is available from Analytical Development Corporation (4405 N. Chestnut Street, Colorado Springs, CO 80907, 719-260-1711).

B

APPENDIX B

Case Study Of Romulus Removal Action

The Comprehensive Environmental Response, Compensation, and Liability Act (CERCLA) Immediate Removal Project at PBM Enterprises of Romulus, Michigan, conducted in 1985, is an excellent example of the application of chemical treatment in a site remediation (Powers 1985). Silver had been recovered from photographic film by a cyanide solution process at the site. Operation ceased in September, 1984. At that time, the site contained 421 tonne (464 ton) of film chips contaminated with up to 1,000 ppm cyanide stored in 17 semi-trailers and two roll-off boxes and about 7,570 L (2,000 gal) of cyanide- or hypochlorite-contaminated liquid wastes. In addition, about 60 m^3 (80 yd^3) of contaminated soil was excavated for treatment. (Cyanide-contaminated soil and surface waters were also found at the site, but these were not part of the emergency removal effort.) Incineration was considered, but was ruled out because of considerations of safety, cost, regulatory, and the public attitude. It was decided to treat the liquids and chips chemically by oxidation with sodium hypochlorite, a method of destroying cyanide that is commonly used in the metal plating and processing industries.

The treatment system was a batch process using sodium hypochlorite in conjunction with pH adjustments to oxidize the sodium cyanide in the following reaction:

$$2NaCN + 5NaOCl + H_2O \rightarrow N_2 + 2NaHCO_3 + 5NaCl \qquad [B.1]$$

The specific process had been developed by Mid-America Services of Riverdale, Illinois, and had been used to treat 7.3 million kg (16 million lb) of cyanide-contaminated film chips in 1984. On-site remediation operations were started on April 11, 1985 and completed on October 8, 1985. Cyanide levels were reduced to less than 20 ppm in all streams except for the contaminated soils in which the posttreatment cyanide level was approximately 60 ppm.

Treatment was conducted in roll-off boxes lined with polyvinyl chloride (PVC) membrane. About 15 m^3 (20 yd^3) of contaminated film chips were loaded into a treatment vessel and then about 13,200 L (3,500 gal) of caustic and hypochlorite solution were added. The mixture was agitated by blowing compressed air through it for two to three hours, and the reaction was monitored. When the reaction was completed, the waste hypochlorite solution was drained into a holding tank and the chips rinsed by adding water to the tank, agitating, and draining. The drained chips were tested for cyanide and, if under 20 ppm, approved for disposal as a nonhazardous waste. The spent hypochlorite solution and rinse liquid were disposed of.

The semitrailers, in which the chips and the drums were stored, were also decontaminated with hypochlorite solution. All the wood was removed from the semitrailers and, along with the drums, decontaminated in the roll-off boxes. The semitrailer shells were vacuum cleaned and then sprayed

Table B.1
Cost Breakdown for Romulus Immediate Removal Action

Item	Cost ($1,000)	Cost, $ per Ton*
Labor	235	469
Travel and Subsistence	39	78
Capital Equipment	213	425
Shipping	48	95
Materials (mainly reagent)	71	142
Sampling and Analysis	3	6
Disposal Costs	45	90
Subcontract Services	28	56
Other	8	16
Total	680	1,358

*Based on 501 tons treated

In addition, EPA performed $124,000 of project management and $3,000 of analytical services in-house or through other government contractors.

(From Powers 1985)

with hypochlorite solution to destroy any residual cyanide. The spent reagent and rinsewater were trucked to a chemical waste treatment facility where it was treated and disposed as a hazardous waste. The treated chips and soil were disposed as nonhazardous wastes.

Problems encountered during the treatment included (1) delivery of poor quality hypochlorite solution, (2) leaks in the PVC liners of field-patched fiberglass vessels, (3) spillage of hypochlorite solution because a plumber inadvertently failed to turn a pump off after a test, and (4) frequent failure of rented air compressors. These problems are not characteristic of chemical treatment operations. They are, however, illustrative of the kinds of problems encountered in the field.

The total cost of the remediation was $698,000, including that for removal and disposal of treated materials and liquid wastes. Additional costs of $127,000 were incurred by EPA for management of the program. While it is difficult to attribute the various cost elements to component treatments, it is reasonable to assume that the vast majority of the cost was associated with treatment of the cyanide-contaminated film chips. The cost of treatment was, therefore, approximately $1,600 per tonne ($1,500 per ton). See table B.1 (on page B.2) for a cost breakdown.

C

APPENDIX C

List Of References

Abdullah M., G. Low, and R. Matthews. 1990. Effects of common inorganic anions on rates of photocatalytic oxidation of organic carbon over illuminated titanium dioxide. *J. Phys. Chem.* 94:6820-5.

Aieta, E.M., E.S. Wong, J. Kuo, and J.M. Montgomery. 1990. Advanced oxidation processes: state of the art review. In *Proc. Am. Chem. Soc. Symposium on Emerging Technologies for Hazardous Waste Treatment.* Atlantic City.

Al-Ekabi H., A. Safarzadeh-Amiri, W. Sifton, and J. Story. 1991. Advanced technology for water purification by heterogeneous photocatalysis. *International J. of Environ. and Poll.* 1(1/2): 125-36.

Alebic-Juretic A., T. Cvitas, and L. Klasinc. 1990. Heterogeneous polycyclic aromatic hydrocarbon degradation with ozone on silica gel carrier. *Env. Sci. and Tech.* 24:62-66.

Andresozzi R., A. Insola, V. Caprio, and M. G. D'Amore. 1991. Ozonation of pyridine in aqueous solution: mechanistic and kinetic aspects. *Water Research* 25:655-9.

Andrews, C.A. 1980. *Photooxidative treatment of TNT contaminated waste waters.* Report No. WQEé/C 80-137. Naval Weapons Support Center.

Ansari, A.S., I.A. Khan, and R. Ali. 1985 UV degradation of arginine in the presence of hydrogen peroxide: involvement of hydroxyl radical in the photolytic process. *J. Radiat. Res.* 26:321-9.

Arumugam, V. 1976. Recovery of chromium from spent chrome tan liquor by chemical precipitation. *Indian J. Environ. Health* 18(1): 47-57.

Baes, C.F., and R.E. Mesmer. 1976. *Hydrolysis of cations.* New York: John Wiley and Sons.

Bailey, F.E. and J.V. Koleske. 1976. *Polyethylene oxide.* New York: Academy Press.

Balasubramanian, D. and B. Chandani. 1983. A poor chemists crown. *J. Chem. Ed.* 60(1):77.

Barber, N.R. 1978. Sodium bicarbonate helps metal plant meet federal standards. *Indus. Wastes* 24(1) :26, 29.

Barr, W.A. 1976. *An evaluation of the engineering design parameters of hydrogen peroxide, UV oxidation of refractories in waste water.* Report No. USNA-EPRD-31. Annapolis, Md.: US Naval Academy.

Barreto R., K. Gray, and K. Anders. 1994. Photocatalytic degradation of methyl-tert-butyl ether in TiO_2 slurries: a proposed reaction scheme. *Water Research.* In review.

Bellamy W.D., B. Langlais, G. Lykins, K. Rakness, C.M. Robson, and P. Schulhof. 1991. Economics of ozone systems: new installations and retrofits. In *Ozone in water treatment: application and engineering,* ed. B. Langlais, D.A. Reckhow, and D.R. Brink, 491-541. Denver: American Water Works Assoc. Research Foundation.

Beltran F., V. Gomez-Serrano, and A. Duran. 1992. Degradation kinetics of p-nitrophenol ozonation in water. *Water Research* 26:9-17.

Benefield, L.D., J.F. Judkins, Jr., and B.L. Weand. 1982. *Process chemistry for water and wastewater treatment.* Englewood Cliffs, N.J.: Prentice-Hall, Inc.

Benitez F., J. Beltran-Heredia, and T. Gonzalez. 1991. Kinetics of the reaction between ozone and MCPA. *Water Research* 25:1345-9.

Bhakta D., S. Shukla, M.S. Chandrasekharalah, J.L. Margrave. 1992. A novel photocatalytic method for detoxification of cyanide wastes. *Env. Sci. Tech.* 26:625-6.

Bhattacharyya, D., A. Jumawan, G. Sun, C. Sund-Hagelberg, and K. Schwitzebel. 1981. Precipitation of sulfide: bench-scale and full-scale experimental results. In *AIChE Sympos. Series, Water-1980* 77(209): 31-42.

Bhattacharyya, D., A.B. Jumawan, and R.B. Grieves. 1979. Separation of toxic heavy metals by sulfide precipitation. *Sep. Sci. Technol.* 14:441-52.

Bhattacharyya, D., J.H. Shin, G.H. Sun, and A.B. Craig, Jr. 1981. Application of sulfide precipitation for the removal of heavy metals from industrial wastewaters. In *Proc. 2nd World Congress of Chem. Engrg.* VI:548-52.

Bowers, A.R., G. Chin, and C.P. Huang. 1981. Predicting the performance of a lime-neutralization/precipitation process for the treatment of some heavy metal-laden industrial wastewaters. In *Proc. 13th Mid-Atlantic Indus. Waste Conf.* 13:51-62.

Brantner, K.A., and E.J. Cichon. 1981. Heavy metals removal: comparison of alternative precipitation processes. In *Proc. 13th Mid-Atlantic Indus. Waste Conf.* 13:43-50.

Bricka, R.M., and M.J. Cullinane, Jr. 1987. Comparative evaluation of heavy metal immobilization using hydroxide and xanthate treatment. In *Proc. 42nd Purdue Indus. Waste Conf.* 42:809-818.

Brown, J.F., M.E. Lynch, J.C. Carnahan, and J. Singleton. 1982. Chemical destruction of PCBs in transformer oil. In *Proc. Detoxification of Hazardous Waste,* 201. ACS. New York. Aug., 1981.

Brunelle, D.J. and D.A. Singleton. 1983. Destruction/removal of PCBs from non-polar media. Reaction of PCB with poly(ethylene glycol)/KOH. *Chemosphere* 12:183.

Brunelle, D.J. and D.A. Singleton. 1985. Chemical reaction of polychlorinated biphenyls on soils with poly(ethylene glycol)/KOH. *Chemosphere* 14(2): 173-181

Bunnett, J.F. 1978. Aromatic substitution by the SRN$_1$, mechanism. *Acc. Chem. Res.* 11:413.

Calvosa L., A. Monteverdi, B. Rindone, and G. Riva. 1991. Ozone oxidation of compounds resistant to biological degradation. *Water Research* 25:985-93.

Camp, D.W. 1991. *Effect of Lamp-Coating Mineral Deposits on UV-Oxidation of Ground Water.* UCRL-JC-107037. Lawrence Livermore National Laboratory. Livermore, Ca.

Chang, T.-K. 1985. *Coprecipitation and adsorption for removal of cadmium, lead, and zinc by the lime-soda ash water softening process.* Ph.D. diss., Purdue University, West Lafayette, Ind.

Chang, T.-K., and R.W. Peters. 1985. Removal of cadmium from contaminated groundwaters through coprecipitation and adsorption in lime softening operations. In *Proc. 17th Mid-Atlantic Indus. Waste Conf.* 17:455-74.

Christian, G.D. 1977. *Analytical Chemistry.* 2d ed. New York: John Wiley and Sons.

Cleland M.R. 1992. High power electron accelerators for industrial radiation processing. In *Radiation processing of polymers*, ed. A. Singh and J. Silverman, 23-42. New York: Hanser Publishers.

Clifford, D., S. Subramonian, and T.J. Sorg. 1986. Removing dissolved inorganic contaminants from water. *Env. Sci. Tech.* 20(11): 1072-80.

Connor, J.R. 1990. *Chemical fixation and solidification of hazardous wastes.* New York: Van Nostrand Reinhold Publishing Co.

Dean, J.A., ed. 1979. *Lange's handbook of chemistry.* 12th ed. New York: McGraw-Hill Book Co.

dePercin, P. 1991. Remedial action, treatment, and disposal of hazardous waste. In *Proc. Seventeenth Annual Hazardous Waste Research Symposium,* 511. RREL, Cincinnati. EPA/600/9-91/002. US EPA. April.

Dickson L.W. and A. Singh. 1986. A review of applications of radiolysis in the adsorbed state. In *Proc. 7th Canadian Nuclear Society Conference.* Toronto.

Dieckmann M.S., K.A. Gray, and P.V. Kamat. 1992. Photocatalyzed degradation of adsorbed nitrophenolic compounds on semiconductor surfaces. *Wat. Sci. Tech.* 25(3):277-9.

Dieckmann M.S. 1993. unpublished results. University of Notre Dame.

Edwards, J.E. and T. Bonham. 1988. Industrial plant expansion: groundwater and soil cleanup. In *Proc. 5th National Conference on Hazardous Wastes and Hazardous Materials.* Las Vegas, Nev.

Farooq S., C. Kurucz, T. Waite, and W. Cooper. 1993. *Disinfection of wastewaters: high energy electron vs. gamma irradiation. Water Research* 27(7): 117-1184.

Faust, S.D., and C.M. Schultz. 1983. The efficacy of removal of heavy metals from water by calcite. *J. Environ. Sci. Health.* A18(1): 95-102.

Federal Remediation Technologies Roundtable. 1992. *Synopses of federal demonstration of innovative site remediation technologies.* 2d ed.

Fender, R.G., A. MacGregor, and K.E. Patterson. 1982. Sulfide precipitation investigation and system design for zinc foundry wastewater. In *Proc. 14th Mid-Atlantic Indus. Waste Conf.* 14:268-277.

Fenton, H.J. 1984. Oxidation of tartaric acid in the presence of iron. *J. Chem. Soc.* 65:899-910.

Ferguson, T.L., and C.J. Rogers. 1990. *Field applications of the KPEG process for treating chlorinated wastes.* Project Officers Report. Order No. Pb 89 212 724/AS and Comprehensive Report on the KPEG Process for Treating Chlorinated Wastes Order No. PB90 163 643. EPA/600/S2-90/026. Risk Reduction Engineering Laboratory. Cincinnati. July.

Flynn, C.M., Jr., T.G. Carnahan, and R.E. Lindstrom. 1980. *Adsorption of heavy metal ions by xanthated sawdust, report of investigations - 8427.* Reno, Nev.: US Bureau of Mines.

Franklin Research Center. 1982. *Technical and cost of hazardous wastes detoxification.* Proposal No. 22. Prepared for US EPA. Philadelphia.

Friedman, A.J. and Y. Halpern. 1992a. Chemical Waste management, Inc. Geneva, Research Center 1950 S. Batavia Avenue, Geneva, IL 60134.

Friedman, A.J. and Y. Halpern. 1992b. *The KGME process.* Patent Pending.

Gerischer H. and A. Heller. 1991. The role of oxygen in photooxidation of organic molecules on semiconductor particles. *J. Phys. Chem.* 95:5261-7.

Gerischer H. and A. Heller. 1992. Photocatalytic oxidation of organic molecules at TiO_2 particles by sunlight in aerated water. *J. Electrochem. Soc.* 139(1): 113-8.

Gold, M.H., H. Kutsuki, and M.A. Morgan. 1983. Oxidative degradation of lignin by photochemical and chemical radical generating systems. *Photochem. Photobiol.* 38:647-51.

GRC, Environmental, Inc. 1992. *Alkaline dechlorination using dimethyl sulfoxide.* East Syracuse, New York. June

Guittonneau, S., J. DeLaat, J.P. Duguet, C. Bonnel, and M. Dore. 1990. Oxidation of parachloronitrobenzene in dilute aqueous solution by O_3 + UV and H_2O_2 + UV: a comparative study. *Ozone Sci. and Eng.* 12:73-94.

Heeks, R.E., L.P. Smith, and R.M. Perry. 1991. Oxidation technologies for groundwater treatment. In *ACS Symposia Series* 468:110-32.

Heindl, A and O. Hutzinger. 1987. Search for industrial sources of PCDD/PCDF, III. Short-chain chlorinated hydrocarbons. *Chemosphere* 16:1949-57.

Heller A. 1987. Industrial aspects of semiconductor photochemistry. *New J. of Chem.* 11(2): 187-9.

Higgins, T.E., and V.E. Slater. 1984. Combined removal of Cr, Cd, and Ni from wastes. *Environ. Prog.* 3(1): 12-25.

Ho, P.C. 1986. Photooxidation of 2,4-d dinitrotoluene in aqueous solution in the presence of hydrogen peroxide. Environ. Sci. Technol. 20(3): 260-7.

Hohman, S.C. 1985. Sulfide precipitation of metals in aqueous systems: selective precipitation and sludge stability. M.S. Thesis, University of Kentucky, Lexington.

Kamaraj, P., S. Jacob, and S. Srinivasan. 1989. Removal of heavy metals from wastewater by sulphide precipitation technique. *Bull. Electrochem.* 5(8): 572-4.

Kamaraj, P., S. Jacob, N. Sathyamurthy, and D. Srinivasan. 1990. Short communication: removal of heavy metals from waste water by sulphide precipitation process. *Indian J. Technol.* 28: 718-20.

Kamaraj, P., S. Jacob, N. Sathyamurthy, and D. Srinivasan. 1991. Sulphide precipitation technique in the removal of heavy metals. *Indian J. Environ. Health* 33(2): 208-12.

Kawamura S. 1991. *Integrated design of water treatment facilities.* New York: John Wiley and Sons, Inc.

Kelly, R.L., W.K. Gauger, and V.J. Srivastava. 1990. Application of Fenton's reagent as a pretreatment step in biological degradation at polyaromatic hydrocarbons. In *Proc. Inst. of Gas Tech. Annual oil, gas, coal and Environmental Biotechnology Symposium.* New Orleans.

Kim, B.M. 1981. Treatment of metal containing wastewater with calcium sulfide. In *AIChE Sympos. Series, Water 1980* 77(209): 39-48.

Kim, B.M., and P.A. Amodeo. 1983. Calcium sulfide process for treatment of metal-containing wastes. *Environ. Prog.* 2(3): 175-80.

Klee, A., C.J. Rogers, and T. Tiernan. 1984. Ind. Environ. Res. Lab. Report, EPA-600/2-84-071. Cincinnati: US EPA. (Avail. NTIS, Order No. PB.84-170059, 75pp.)

Kolthoff, I.M. 1932. Theory of coprecipitation - the formation and properties of crystalline precipitates. *J. Phys. Chem.* 36(3): 860.

Kornel, A. and C.J. Rogers. 1987. *Chemical destruction of halogenated aliphatic hydrocarbons, reacting with an alkali metal glycolate.* U.S. Patent 4,675,464.

Kornel, A., Rogers, C.J., and H.L. Sparks. 1991a. *Method for the destruction of halogenated organic compounds in a contaminated medium, polyethylene glycol with alkali metal hydroxide.* U.S. Patent No. 5,019,175. May 28

Kornel, A., Rogers, C.J., and H.L. Sparks. 1991b. *Method for the destruction of halogenated organic compounds in a contaminated medium, adding alkali metal carbonates.* U.S. Patent No. 5,039,350. Aug. 13.

Kornel, A., Rogers, C.J., and H.L. Sparks. 1991c. Method for the base catalyzed decomposition of halogenated and nonhalogenated organic compounds in a contaminated medium. U.S. Patent No. 5,064,526. Nov. 12.

Kotronarou A., G. Mills, and M.R. Hoffmann. 1991. Ultrasonic irradiation of p-nitrophenol in aqueous solution. *J. Phys. Chem.* 95:3630-8.

Kotronarou A., G. Mills, and M.R. Hoffmann. 1992. Decomposition of parathion in aqueous solution by ultrasonic irradiation. *Env. Sci. Tech.* 26:1460-2.

Ku, Y. 1982. Sulfide precipitation of heavy metals: development of reaction equilibrium model and establishment of chelating agents effect on precipitation. M.S. Thesis, University of Kentucky, Lexington.

Ku, Y. 1986. Removal of heavy metals by sulfide precipitation in the presence of complexing agents. Ph.D. diss., Purdue University, West Lafayette, Ind.

Ku, Y., and R.W. Peters. 1986. The effect of weak chelating agents on the removal of heavy metals by precipitation processes. *Environ. Prog.* 5(3): 147-53.

Ku, Y., and R.W. Peters. 1988. The effect of complexing agents on the precipitation and removal of copper and nickel from solution. *Particulate Sci. and Technol.* 6(4): 441-66.

Kuo, W.G. 1992. Decolorizing dye wastewater with Fenton's reagent. *Water Research* 26:881-6.

Kurucz C., T. Waite, W. Cooper, and M. Nickelsen. 1990. Full-scale electron beam treatment of hazardous wastes-effectiveness and costs. In *Proc. 45th Industrial Waste Conference*. Chelsea, Mich.: Lewis Publishers.

Kusakabe K., S. Aso, T. Wada, J. Hayashi, S. Morooka, and K. Isomura. 1991. Destruction rate of volatile organochlorine compounds in water by ozonation with ultraviolet radiation. *Water Research* 25(10): 1199-203.

Langlais B., D.A. Reckhow, and D.R. Brink. 1991. *Ozone in water treatment: application and engineering*. Denver: American Water Works Assoc. Research Foundation.

Leckie, J.O., D.T. Merrill, and W. Chow. 1985. Trace element removal from power plant waste streams by adsorption/ coprecipitation with amorphous iron oxyhydroxide. In *AIChE Sympos. Series, Separation of Heavy Metals and Other Trace Contaminants* 81(243): 28-42.

Legube B. 1985. Identification of a few organics and attempted quantification upon disinfection with ozone of a biologically treated wastewater. In *Proc. Intl. Conf, The Role of Ozone in Water and Wastewater Treatment*, ed. R. Perry and R. E. McIntyre. Selper, LTD. London.

Legube B. 1986. Ozonation of naphthalene in aqueous solution a) Part 1: Ozone consumption and ozonation products; b) Part 2: Kinetics studies of the initial reaction step. *Water Research* 20:197.

Lowry T.H. and K.S. Richardson. 1981. *Mechanism and theory in organic chemistry*. 2d ed. New York: Harper and Row Publishers.

Macomber, R.S., M. Orchin, G. Garrett, and H. Braus. 1983. In-house report on KPEG technology fields of new KPEG Reactor on PCDD contaminated soils. Contract No. 68-03-2846 EPA, Modification, No. 46. University of Cincinnati, Ohio. Feb. 24, 1987.

Magrini K., R. Goggin, A. Watt, and A. Taylor. 1992. Water composition effects on the photocatalytic oxidation of aqueous trichloroethylene. Paper presented at Spring Meeting of the American Chemical Society, April, in San Francisco.

Manilal V., A. Haridas, R. Alexander, and G. Surender. 1992. Photocatalytic treatment of toxic organics in wastewater: toxicity of photodegradation products. *Wat. Res.* 26(8): 1035-8.

Maruyama, T., S.A. Hannah, and J.M. Cohen. 1975. Metal removal by physical and chemical treatment processes. *J. Water Pollut. Contr. Fed.* 47(5): 962-75.

Matsuzuru, H., M. Toshikuni, A. Yamanaka, and N. Moriyama. 1982. *Oxidative degradation of powdered cation exchange resin by FeII. Catalyzed hydrogen peroxide* 23. Report Number JAERI-M-82-087. Japan Atomic Energy Institute.

Matthews R.W. 1987. Solar-electric water purification using photocatalytic oxidation with TiO_2 as a stationary phase. *Solar Energy* 38(6): 405-13.

Matthews R.W. 1988. Kinetics of photocatalytic oxidation of organic solutes over titanium dioxide. *J. of Catalysis* 111:264-72.

McAnally, S.L., L. Benefield, and R.B. Reed. 1984. Nickel removal from a synthetic nickel-plating wastewater using sulfide and carbonate for precipitation and coprecipitation. *Sep. Sci. Technol.* 192(3): 191-217.

McFadden, F., L. Benefield, and R.B. Reed. 1985. Nickel removal from nickel plating wastewater using iron, carbonate, and polymers for precipitation and co-precipitation. In *Proc. 40th Purdue Indus. Waste Conf.* 40:417-29.

J.M. Montgomery Consulting Engineers. 1985. *Water treatment principles and design*. New York: John Wiley and Sons, Inc.

Neuman, R. and Y. Sasson. 1983. Mechanism of base catalyzed reactions in phase-transfer systems with polyethylene glycol as catalysts. *Tetrahedron* 20:3734-37.

Newkirk, D.D., M.G. Warner, and S. Barros. 1981. Treatability studies on heavy metal removal in selected inorganic chemical industries. In *Proc. 36th Purdue Indus. Waste Conf.* 36:17-28.

Nickelsen M., W. Cooper, C. Kurucz, and T. Waite. 1992. Removal of benzene and selected alkyl-substituted benzenes from aqueous solution utilizing continuous high-energy electron irradiation. *Env. Sci. Tech.* 26:144-52.

Ogata, Y., K. Tomizawa, and K. Takagi. 1981. Photooxidation of formic, acetic and proprionic acids with aqueous hydrogen peroxide. *Canadian J. of Chemistry* 59:14.

Ollis D.F., E. Pelizzetti and N. Serpone. 1989. Heterogeneous photocatalysis in the environment: application to water purification. In *Photocatalysis - fundamentals and applications,* ed. N. Serpone and E. Pelizzetti, 603-637. New York: Wiley.

Ollis, D. 1985. Contaminant degradation in water. *Env. Sci. Tech.* 19(6): 480-4.

Osaki, S., S. Sugihara, and T. Kaji. 1990. Treatment of radioactive waste phenol with Fenton's oxidation. *Radioisotopes.* 39:174-7.

Patterson, J.W. 1988. *Metal treatment and recovery.* In *Metal speciation: theory, analysis, and application*, ed. J.R Kramer and H.E. Allen, 333-345. Chelsea, Mich: Lewis Publishers, Inc.

Patterson, J.W., and R.A. Minear. 1975. *Physical-chemical methods of heavy metals removal.* In *Heavy metals in the aquatic environment*, ed. P.A. Krenkel, 261-276. Oxford, England: Pergamon Press.

Patterson, J.W., H.E. Allen, and J.J. Scala. 1977. Carbonate precipitation for heavy metals pollutants. *J. Water Pollut. Control Fed.* 49(12): 2397-410.

Pelizzetti E., C. Minero and V. Maurino. 1990. *Adv. Colloid Interface Sci.* 32:271-316.

Perry, J.H., C.H. Chilton, and S.D. Kirkpatrick. 1963. *Chemical engineers' handbook.* New York: McGraw Hill Book Company.

Peters, R.W., and T.-K. Chang. 1984. Removal of heavy metals by coprecipitation and adsorption on the lime-soda ash water softening process. In *Proc. 15th Annual Meeting of the Fine Particle Society.* Orlando, Fla. August 19-22.

Peters, R.W, and T.-K. Chang. 1985. The effect of particle size distribution and morphology of $CaCO_3$ precipitation in the presence of Zn and Cd. In *Proc. 16th Annual Meeting of the Fine Particle Society.* Miami Beach, Fla. April 22-26.

Peters, R.W., and Y. Ku. 1984. Removal of heavy metals from industrial plating wastewaters by sulfide precipitation, 279-311. In *Proc. Industrial Wastes Symposia, 57th Water Pollution Control Federation Annual Conference.*

Peters, R.W., and Y. Ku. 1985. Batch precipitation studies for heavy metal removal by sulfide precipitation. In *AIChE Sympos. Series, Separation of Heavy Metals and Other Contaminants* 81(243): 9-27.

Peters, R.W., and Y. Ku. 1987. *The effect of citrate, a weak complexing agent, on the removal of heavy metals by sulfide precipitation*, 147-169. In *Metals speciation, separation, and recovery*, ed. J.W. Patterson and R. Passino. Chelsea, Mich.: Lewis Publishers, Inc.

Peters, R.W., and Y. Ku. 1988. The effect of tartrate, a weak complexing agent, on the removal of heavy metals by sulfide and hydroxide precipitation. *Particulate Sci. and Technol.* 6(4): 421-39.

Peters, R.W., E. Eriksen, and Y. Ku. 1985. Segregated removal of heavy metal species from mixed-metal plating wastewaters by selective precipitation. In *Proc. 1985 Triangle Conference on Environmental Technology.* Raleigh, N.C. April 3-4.

Peters, R.W., Y. Ku, and D. Bhattacharyya. 1984. The effect of chelating agents on the removal of heavy metals by sulfide precipitation. In *Proc. 16th Mid-Atlantic Indus. Waste Conf.* 16:289-317.

Peters, R.W., Y. Ku, and D. Bhattacharyya. 1985. Evaluation of recent treatment techniques for removal of heavy metals from industrial wastewaters. In *AIChE Sympos. Series, Separation of Heavy Metals and Other Contaminants* 81(243): 165-203.

Peters, R.W., Y. Ku, D. Bhattacharyya, and L.-F. Chen. 1984. Crystal size distribution of sulfide precipitation of heavy metals, industrial crystallization, 84:111-23. In *Proc. 9th Sympos. Indus. Crystal.* The Hague, Netherlands. September 25-28.

Peterson, R.L. 1986. *Method for decontaminating soil.* U.S. Patent No. 4,574,013. March 4.

Pignatello, J.J. 1992. Dark and photo-assisted Fe3+-catalyzed degradation of chlorophenoxy herbicides by hydrogen peroxide. *Env. Sci. and Tech.* 26:944-51.

Powers, R.E. 1985. *On scene coordinator's report, CERCLA immediate removal project.* Romulus, Mich.: PBM Enterprises. Dec. 16.

Pugsley, E.B., C.Y. Cheng, D.M. Updegraff, and L.W. Ross. 1970. Removal of heavy metals from mine drainage in Colorado by precipitation. *Chem. Engrg. Prog. Sympos. Series, Water - 1970* 67(107): 75-89.

Purbaugh, T. 1992. Telephone conversation with J. Verbicky. June 6.

Rabosky, J.G., and T. Altares, Jr. 1983. Wastewater treatment of a small chrome plating shop: a case history. In *Proc. 38th Purdue Indus. Waste Conf.* 38:449-56.

Reynolds G. 1989. Aqueous ozonation of pesticides: a review. *Ozone Sci. Engrg*. 11:339.

Rice, R.G. 1981. Ozone for the treatment of hazardous materials. In *Proc. WATER-1980 Symposium Series*. American Institute of Chemical Engineers. 29(77): 79-107.

RadTech. 1992. Proceedings of RadTech North America Conference.

Rogers, C.J. 1983. Incineration treatment of hazardous waste. In *Proc. 8th Annual Research Syymposium,* 197-201. EPA600/9 83-003, PB83-210450. Res. Dev. (Rep.). Cincinnati: US EPA.

Rowland, M.A. 1989. Groundwater treatment with ultraviolet light and hydrogen peroxide. In *Proc. National Water Well Association Outdoor Action Conference*, 659-672.

Salutsky, M.L. 1959. Precipitates: their formation, properties, and purity. In *Treatise on analytical chemistry,* ed. I.M. Kolthoff and P.J. Elving, Part I, Vol. I, Sec. B. New York: Interscience.

Sawyer, C.N., and P.L. McCarty. 1978. *Chemistry for environmental engineering*. 3d ed. New York: McGraw-Hill Book Co.

Schiavello, M., ed. 1988. *Photocatalysis and environment.* Dortrecht: Kluwer.

Schlauch, R.M., and A.C. Epstein. 1977. *Treatment of metal finishing wastes by sulfide precipitation.* EPA Report. EPA 600/2-77-049. Cincinnati: US EPA

Scott, M.C. 1979. An EPA demonstration plant for heavy metals removal by sulfide precipitation, 126. In *Proc. 2nd Conf. Adv. Pollut. Contr. for Metal Fin. Indus.* EPA 600/8-79-014.

Sedlak, D.L. and A.W. Andren. 1991. Oxidation of chlorobenzene with Fenton's reagent. *Env. Sci. and Tech.* 25:777-82.

Sehested K., H. Corfitzen, J. Holcman, C.H. Fischer, and E.J. Hart. 1991. The primary reaction in the decomposition of ozone in acidic aqueous solutions. *Env. Sci. and Tech.* 25:1589-96.

Sheffield, C.W. 1981. Treatment of heavy metals at small electroplating plants. In *Proc. 36th Purdue Indus. Waste Conf.* 36:485-92.

Sillen, L.G., and A.E. Martell. 1971. *Stability constants of metal ion complexes*. London: Chemical Society.

Singh A. 1986. Chemical and biochemical aspects of activated oxygen: singlet oxygen, superoxide anion, and related species. In *Handbook of free radicals and antioxidants in biomedicine*, ed. J. Miguel, A.T. Quintanilha, and H. Weber, Vol. 1, 17-27. CRC Press.

Singh A., N. Sagert, J. Borsa, H. Singh, and G. Bennett. 1985. The use of high-energy radiation for the treatment of wastewater: a review. In *Proc. 8th Wastewater Treatment Conference, Environment.* Montreal, Canada. Nov.

Singh A., W. Kremers, P. Smalley, and G. Bennett. 1985. Radiolytic dechlorination of polychlorinated biphenyls. *Radiat. Phys. Chem.* 25(1): 11-9.

Smith, D.K. 1988. *Application of UV lamp technology for water treatment or sterilization. By-products and environmental impact.* Report No. 647 U 588. Canadian Electrical Association.

Smith, J.F. and G.L. Bubbar. 1979. *The chemical destruction of polychlorinated biphenyls by sodium naphthalenide.* Report from the Guelph-Waterloo Centre for Graduate Work in Chemistry, Department of Chemistry, University of Waterloo. Waterloo, Ontario: University of Waterloo.

Smith, R.M., and A.E. Martell. 1976. *Critical stability constants.* New York: Plenum Press, Inc.

Snoeyink, V.L., and D. Jenkins. 1980. *Water chemistry.* New York: John Wiley and Sons, Inc.

Somich C.J., M.T. Muldoon, and P.C. Dearney. 1990. On-site treatment of pesticide waste and rinsate using ozone and biologically active soil. *Env. Sci. and Tech.* 24:745-9.

Sorg, T.J. 1979. Treatment technology to meet the interim primary drinking water regulations for organics: part 4. *J. Am. Water Works Assoc.* 71(8): 454-66.

Sorg, T.J., M. Csanady, and G.S. Logsdon. 1978. Treatment technology to meet the interim primary drinking water regulations for inorganics: part 3. *J. Am. Water Works Assoc.* 70(12): 680-91.

Stafford U., K.A. Gray, P. Kamat, and A. Varma. 1993. An in-situ diffuse reflectance FTIR investigation of photocatalytic degradation of 4-chlorophenol on a TiO_2 powder surface. *Chem. Phys.* 205(1): 55-61.

Stowell J. and Jensen J. 1991. Dechlorination of chlorendic acid with ozone. *Water Research* 25(1) 83-90.

Sundstrom, D.W. and H.E. Klei. 1986. *Destruction of hazardous compounds by ultraviolet catalyzed oxidation with hydrogen peroxide.* Report #PB87-149357. NTIS.

Taciuk, W. 1979. *Process for thermal cracking a heavy hydrocarbon.* U.S. Patent 4,180,455. Dec. 25.

Taciuk, W. 1981a. *Process for recovery of hydrocarbons from inorganic host materials comprising oil sands containing clay and bitumen.* U.S. Patent 4,280,879. July 28.

Taciuk, W. 1981b. *Apparatus and process for recovery of hydrocarbons from inorganic host materials.* U.S. Patent 4,306,961. Dec. 22.

Talbot, R.S. 1984. Co-precipitation of heavy metals with soluble sulfides using statistics for process control. In *Proc. 16th Mid-Atlantic Indus. Waste Conf.* 16:279-88.

Tate, B. 1991. New oxidation process scrubs contaminated soil. *US Army Corps of Engineers, Engineer Update.* July.

Turchi C. speaker. 1992. Pilot-scale study of the solar detoxification of VOC contaminated groundwater. Paper presented at *1992 Summer Meeting of the American Institute of Chemical Engineers.* Minneapolis.

Turchi C.S. and D.F. Ollis. 1990. Photocatalytic degradation of organic water contaminants: mechanisms involving hydroxyl radical attack. *J. of Catalysis* 122:178-92.

Ultrox Process Brochures. 2435 South Anne Street, Santa Ana, Ca.

US EPA. 1978. *Manual of treatment techniques for meeting the interim primary drinking water regulations.* EPA 600/8-77-005.

US EPA. 1980. *Summary report: control and treatment technology for the metal finishing industry; sulfide precipitation.* EPA 625/8-80-003.

US EPA. 1989. *Technology evaluation report and the applications analysis report.* Superfund innovative technology evaluation program. EPA/540/5-89/012, EPA/540/A5-89/012. Ultrox International.

US EPA. 1990. *Handbook on in-situ treatment of hazardous waste contaminated soils,* 22-23. EPA/540/2-90/002.

US EPA. 1992a. *Alternative Treatment Technology Information Center (ATTIC)* (data base). Office of Solid Waste and Emergency Response, Technology Innovation Office. Washington, D.C.

US EPA. 1992b. *Synopses of federal demonstrations of innovative site remediation technologies* 58. Federal Remediation Technologies Roundtable Report. Contract No. 68-W2-004.

US EPA. 1992c. *SoilTech Anaerobic Thermal Processor, Outboard Marine Corporation site.* EPA 540/MR-92/078

US EPA. 1992d. *Technology evaluation report, SoilTech/Anaerobic thermal processor technology.* EPA/540/XX-XX/XXX. SITE Demonstration Bulletin, RREL Draft Report.

Vasilenko, I.I. and A.N. Fedosoua. 1987. Liquid phase heterocatalytic oxidation of phenol by hydrogen peroxide on magnetite. *Zhurnal Prikladnoi Khimii* 60(4): 870-3.

Walling, C. 1975. Fenton's reagent revisited. *Acc. Chem. Res.* 8:125-31.

Watts, R.J., B.W. Tyre, and G.C. Miller. 1991. Treatment of four biorefractory contaminants in soils using catalyzed hydrogen peroxide. *J. Environ. Qual.* 20:832-8.

Weitzman, L. 1982. Treatment and destruction of polychlorinated biphenyls and PCB-contaminated materials. In *Proc. Detoxification of Hazardous Waste,* 131. ACS. New York. Aug., 1981.

Wekhof A. 1991. Treatment of contaminated water, air and soil with UV flashlamps. *Env. Progress* 10(4): 241-7.

Wentsel, R.S., S. Sommerer, and J.F. Kitchens. 1981. *Engineering and development support of general decon technology for the DARCOM installation restoration program, task 2. Literature review on treatment of contaminated lagoon sediment- phase I.* Report under US Army Toxic and Hazardous Materials Agency contract No. DAAK11-80-C-0027. Alexandria, Va.: Atlantic Research Corporation.

Whang, J.S., D. Young, and M. Pressman. 1981. Design of soluble sulfide precipitation system for heavy metals removal. In *Proc. 13th Mid-Atlantic Indus. Waste Conf.* 13:63-71.

Wing, R.E. 1974. Heavy metal removal from wastewater with starch xanthate. In *Proc. 29th Purdue Indus. Waste Conf.* 29:348-56.

Wing, R.E., and W.E. Rayford. 1976. Starch-based products effective in heavy metal removal. In *Proc. 31st Purdue Indus. Waste Conf.* 31:1068-79.

Wing, R.E., and W.E. Rayford. 1977. Heavy metal removal processes for plating rinse waters. In *Proc. 30th Purdue Indus. Waste Conf.* 30:838-52.

Wing, R.E., L.L. NaVickis, B.K. Jusberg, and W.E. Rayford. 1978. *Removal of heavy metals from industrial wastewaters using insoluble starch xanthates.* EPA 600/2-78-085. Cincinnati: US EPA, Office of Research and Development.

Wing, R.E., W.N. Doane, and C.R. Russell. 1975. Insoluble starch xanthate: use in heavy metal removal. *J. Appl. Polym. Sci.* 19:.

Yanagida, S., K. Takahashi, and M. Okahara. 1977. Metal ion complexation of noncyclic poly(oxyethylene) derivatives. Part IV para transfer catalysis of poly(oxyethylene) dimethyl ethers. *Bull. Chem. Soc. Japan* 50:1386.

Yao, C.C.D. and W.R. Haag. 1991. Rate constants for direct reactions of ozone with several drinking water contaminants. *Water Research* 25:761-73.

Zepp, R.G., B.C. Faust, and J. Hoigne. 1992. Hydroxyl radical formation in aqueous reactions pH 3-8, of iron II with hydrogen peroxide: the Photo-Fenton Reaction. *Environmental Science and Technology* 26:313-9.

D

APPENDIX D

Suggested Reading List

Exner, J.H., ed. 1981. *Detoxification of hazardous waste.* Ann Arbor, Mich.: Ann Arbor Science.

US EPA. 1991. *Bibliography of federal reports and publications describing alternative and innovative treatment technologies for corrective action and site remediation.* Federal Remediation Technologies Roundtable. EPA/540/8-91/007. Office of Solid Waste and Emergency Response, Technology Innovation Office. Washington, D.C. May.

US EPA. 1991. *Innovative treatment technologies, overview and guide to information sources.* EPA/540/2-91/002. Technology Innovation Office, Office of Solid Waste and Emergency Response. Washington, D.C. Oct.

US EPA. 1991. *Innovative treatment technologies, semi-annual status report,* No. 2. EPA/540/2-91/001. Technology Innovation Office, Office of Solid Waste and Emergency Response. Washington, D.C. Sept.

US EPA. 1991. *Selected alternative and innovative treatment technologies for corrective action and site remediation a bibliography of EPA information resources, fall update.* EPA/540/8-91/092. Technology Innovation Office, Office of Solid Waste and Emergency Response. Washington, D.C. Nov.

US EPA. 1991. *Synopses of federal demonstrations of innovative site remediation technologies.* Federal Remediation Technologies Roundtable. EPA/540/8-91/009. Washington, D.C. May.

US EPA. 1991. Papers presented at Third Forum on Innovative Hazardous Waste Treatment Technologies: Domestic and International. Dallas. June 11-13. EPA/540/2-91/015. Office of Solid Waste and Emergency Response, Technology Innovation Office and Risk Reduction Engineering Laboratory, Cincinnati. Washington, D.C. Sept.

US EPA. 1992. *Literature survey of innovative technologies for hazardous waste site remediation 1987—1991*. Preliminary Draft. Office of Solid Waste and Emergency Response, Technology Innovation Office. Washington, D.C. Feb.